Reigning the River

New Ecologies
for the Twenty-First Century

Series Editors

ARTURO ESCOBAR,
University of North Carolina, Chapel Hill

DIANNE ROCHELEAU,
Clark University

ANNE M. RADEMACHER

Reigning the River

URBAN ECOLOGIES AND

POLITICAL TRANSFORMATION

IN KATHMANDU

Foreword by Dianne Rocheleau

Duke University Press

DURHAM AND LONDON 2011

© 2011 Duke University Press
All rights reserved
Printed in the United States of America
on acid-free paper ∞
Designed by C. H. Westmoreland
Typeset in Arno Pro by Tseng Information Systems, Inc.
Library of Congress Cataloging-in-Publication Data
appear on the last printed page of this book.

Frontispiece: Cultural heritage activist Huta Ram Baidya at the Bishnumati River. Photo by the author.

For Nancy and Manju

Contents

About the Series viii
Foreword ix
Acknowledgments xiii

Introduction. A Riverscape Undone 1
1. Creating Nepal in the Kathmandu Valley 42
2. Knowing the Problem 57
3. War, Emergency, and an Unsettled City 91
4. Emergency Ecology and the Order of Renewal 116
5. Ecologies of Invasion 139
6. Local Rivers, Global Reaches 155
Conclusion. Anticipating Restoration 175

Notes 185
References 211
Index 237

About the Series

This series addresses two trends: critical conversations in academic fields about nature, sustainability, globalization, and culture, including constructive engagements between researchers within the natural, social, and human sciences; and intellectual and political conversations among those in social movements and other nonacademic knowledge producers about alternative practices and socionatural worlds. The objective of the series is to establish a synergy between these theoretical and political developments in both academic and nonacademic arenas. This synergy is a sine qua non for new thinking about the real promise of emergent ecologies. The series includes works that envision more lasting and just ways of being-in-place and being-in-networks with a diversity of humans and other living and nonliving beings.

New Ecologies for the Twenty-First Century aims to promote a dialogue between those who are transforming the understanding of the relationship between nature and culture. The series revisits existing fields such as environmental history, historical ecology, environmental anthropology, ecological economics, and cultural and political ecology. It addresses emerging tendencies, such as the use of complexity theory to rethink a range of questions on the nature–culture axis. It also deals with epistemological and ontological concerns, in order to build bridges between the various forms of knowing and ways of being that are embedded in the multiplicity of practices of social actors worldwide. This series hopes to foster convergences among differently located actors and to provide a forum for authors and readers to widen the fields of theoretical inquiry, professional practice, and social struggles that characterize the current environmental arena.

Foreword

ANNE RADEMACHER'S TALE of braided social and environmental change in the new ecologies of Kathmandu contributes a uniquely situated urban perspective to New Ecologies for the Twenty-First Century. Her work places several challenging questions and reflections on the table for those who presume to construct such ecologies through modernist development strategies, whether from "above" or "below." *Reigning the River* addresses theoretical and policy issues of development, urban ecology, political ecology, and cultural dimensions of nature within processes of urban and metropolitan growth. The first-person narrative of experience and observation in Kathmandu for over a decade is impressive, weaving political crises in Nepal during the 1990s and early 2000s into a story of ecologically focused activism and vice versa. The Monarchist scandals, the Maoist insurgency, and a massive international development establishment, along with a political revival of Hindu religious tradition and a growing flood of immigrants to the city's riverside, all play a part in the re-making of the river. The Bagmati River is simultaneously sacred to multiple groups as well as constituting a necessary resource for several others with distinct visions of the river's past, present, and future.

Given the status of Nepal as a major focus of international donor attention and funding for decades, the development problems, foibles, structural traps, and lessons woven into this story are relevant to many other places and circumstances. While I initially found this work to be an important contribution to the academic literature across multiple fields, on a second reading I found tremendous resonance with my own observations and experiences in Kenya, the Dominican Republic, Mexico, and the United States. I expect to cite this work with some frequency in upcoming publications about the damage wrought by modernist development and conservation megaprojects currently underway in Mexico and throughout Central America. I can also see the key points from this story playing out in the sites in New York City where I take my students each year to study politics, power, and environmental justice in urban ecologies. As a professor of both environmental studies and metropolitan studies, Anne Rademacher has brought together the environmental history and political ecology of Kathmandu in a way that draws deep and broadly relevant insights from a specific urban case study.

The story of the way that the Bagmati and Bishnumati rivers emerge as actors in this text, along with the people whom the author knows and works with, kept me reading and wanting to know the outcomes. The Bagmati River and its tributaries become real characters, along with various social movements, social actors, and political constituencies. There is a sense of immersion created by the first-person narrative and I can see using it to help American and European students to take in the complexity of Kathmandu in a specific context of massive sustainable development investment and intervention. I would like to know more about the fate of the people evicted from the riverbanks, as well as about some of the other individuals, organizations, and sacred sites described in the text. Rademacher includes a recent update in the last chapter and conclusion and notes her own intent to follow some people and stories into the future. The fact that I am left this curious after reading the text suggests to me that this book is an excellent possibility for encouraging both political and classroom discussion and has a very strong potential for classroom use by upper-level undergraduates and graduate students. I can also imagine this book being read by international environment and development professionals interested in the social, political, and cultural dimensions of urban sustainable development projects and policies.

The theoretical contribution is in the detailed description as well as the interpretation of variously situated actors and their simultaneous embrace of specific modern and traditional takes on the ecology, geography, and culture of Kathmandu in the past, present, and future. The entangled web of actors, policies, and actions focused on the rivers and urban ecology of Kathmandu is woven through the story of the monarchy as it wends its way to near erasure, and through the story of the rise of multiple and contested alternative forms of authority and power over the city, the people, and the rivers. The role of religion and "tradition" in the negotiation of sustainable development politics is a pivotal point of interest in the unfolding story of Kathmandu and its rivers. Likewise, the work of engineers and architects plays into various scenarios of worsening water quality and erratic flow. The spectacle of celebrities and politicians rafting down the river in the midst of fetid odors, pomp and ceremony, and media frenzy demonstrates the merging of sustainable development economics and politics in the twenty-first century.

Anne Rademacher's book is readable, informative, and theoretically and empirically compelling. *Reigning the River* will be of interest to scholars and students of Asian studies, international urban and metropolitan studies, Nepali studies, urban ecologies, urban political ecologies, political ecology (anthropology, geography, and sociology), and environmental studies. The book also addresses the theoretical and policy concerns of people working at the interface of environmental science, science and technology studies, and development studies. This is an intersection that academics, policymakers, and social movements will have to visit frequently in order to make sense of a rapidly urbanizing world that includes complex and power-laden entanglements between urban and rural spaces and peoples and ecosystems across the planet and over time.

DIANNE ROCHELEAU
Professor of Geography and
Global Environmental Studies
Clark University
Worcester, Massachusetts
Series Coeditor of New Ecologies
for the Twenty-First Century

Acknowledgments

THE SCHOLARLY JOURNEY recounted here spans many years and has been enriched by crossing many lives. This book is first and foremost a product of gifts—of time, conversation, and insight, as well as institutional and material support. I owe great intellectual and personal debts to those whose contributions, small and large, made this project possible.

My gratitude reaches back to 1991, when I first visited Nepal as an undergraduate. In the winter of that year, I joined a Carleton College Semester Abroad program that inspired my interests, curiosities, and commitment. I am grateful to Professor James Fisher, who led that program, and for the care with which he introduced a very lucky group of students to the rich histories, languages, and cultures of Nepal. It was also in the winter of 1991 that I first stayed in the home of a family that would continue to welcome me into their household over the nearly two decades to follow. I am grateful to them beyond words for their warmth, generosity, and enduring friendship. In many ways, this book is for them.

I am also deeply grateful to the many friends, colleagues, and informants in Kathmandu who contributed to this study. Geeta Manandhar, Shambhu Ojha, Banu Ojha, and Sama Adhikari were patient, careful, and rigorous teachers of Nepali over my years of fieldwork. The staff of Lumanti Support Group for Shelter, particularly Lajana Manandhar and Sama Vajracharya, offered essential assistance by providing access to the Lumanti Research Center and by teaching me about the riparian *sukumbasi* (landless migrant) communities on the Bagmati and Bishnumati rivers. Huta Ram Baidya gave generously of his time and wisdom, guiding me to understand his ideas and aspirations during countless walks along the riverscape. I would also like to thank several organizations for their cooperation with my research and endless inquiries. These include IUCN-Nepal, ICIMOD, SAGUN, SEARCH, Friends of the Bagmati, the Asian Development Bank, USAID, John Sanday Associates, the Asia Foundation, KEEP,

xiv   *Acknowledgments*

and ECCA. I am grateful to Sama Adhikari, Bikash Dhital, Dorje Guring, and Jeevan Raj Sharma for the invaluable field research assistance that they provided at various points in the project.

I also wish to acknowledge the varied contributions of Nishi and Bhim Adhikari, the Dhittal family, Anne Kaufman, Linda Kentro, Laura Kunreuther, Shambhu and Sunita Pradhan, Prafulla Man Pradhan, Laxman Shrestha, Umesh and Karuna Shrestha, Rajiv Sinha, Haydie and David Sowerwine, Deepak Tamang, and Mukta Tamang. Although political circumstances in Nepal prevent me from safely naming them in this volume, I acknowledge with gratitude the residents of and advocates for riparian communities on the banks of the Bagmati and Bishnumati; these residents gave generously of their time, their stories, and their confidence over the course of this project.

While conducting field research and writing this book, I had the opportunity to work with, and learn from, an extraordinary group of mentors and colleagues. I owe an enormous debt to the environmental anthropology community at Yale University, which in my very rich experience there gave me the privilege of learning from Michael Dove, Carol Carpenter, Stacy Leigh Pigg, Helen Siu, and Eric Worby. Helen Siu and Michael Dove were extraordinary, inspiring mentors whose rigorous but always supportive engagement with my work left indelible imprints on my scholarship and my aspirations as a teacher. I also wish to acknowledge with deep gratitude the contributions of Arun Agrawal, Steven Andrews, Elisabeth Barsa, Ken Bauer, Amita Baviskar, Manu Bhagavan, William Burch, Seth Cook, Sienna Craig, James Fisher, Eva Garen, Gordon Geballe, David Graeber, Tika Gurung, Shubhra Gururani, Susan Hangen, Sondra Hausner, John Hendrickson, Kathe Goria-Hendrickson, David Holmberg, Aban Marker Kabraji, Sarah Koch-Schulte, Laura Kunreuther, Kathryn March, Andrew Mathews, Leah Mayor, Raj Patel, Mieka Ritsema, Tom Robertson, Kai Schafft, Sara Shneiderman, K. Sivaramakrishnan, Mark Turin, Laurie Vasily, Vron Ware, Eva Cuadrado Worden, Scarlet Wynns, Laura Meitzner Yoder, Abraham Zablocki, Eleanor Zelliot, and my many friends and colleagues from the Yale Program in Agrarian Studies and the Cornell University Nepal Studies Association.

This book took shape through critical engagements with my colleagues at New York University. I am particularly grateful to the faculties in the metropolitan studies program and the environmental studies program, whose unfailing support helped me to more fully synthesize and analyze the research. I extend sincerest thanks to Neil Brenner, Harvey Molotch, Caitlin Zaloom, and Danny Walkowitz in particular; without their initial faith in the work, especially over my two-year tenure as a faculty fellow with the NYU metropolitan studies program, writing the book may not have been possible. I am especially grateful to Harvey Molotch for suggesting the book title. Likewise, colleagues in the environmental studies program, most notably Dale Jamieson, Chris Schlottmann, and Tyler Volk, provided productive intellectual engagement, unfailing support, and essential companionship. I am grateful as well to the inspiring faculty and exceptional staff of the department of social and cultural analysis, the seemingly superhuman Amanda Amjun and Zahra Ali, and the rich network of social science and environmental studies colleagues that forms the very fertile ground in which environmental scholarship grows at NYU.

The insights and input of many scholars and practitioners enriched various elements of this work, and to each I am extremely grateful. Although the list is longer than I can name here, I wish to mention in particular my collaborator and cherished friend Nikhil Anand, the tireless members of the research network on urban ecologies in South Asia, and the endlessly energetic and inspiring Aban Marker Kabraji. Tejaswini Ganti and Noelle Stout kept me on track through productive and insightful reflections on my work, and the patient encouragement of Julie Elman and Thuy Linh Tu saw me through its very last stages. Drs. K. Sivaramakrishnan and Helen Siu have supported me throughout the long journey from their classrooms to my own; I have at all stages been extremely grateful for their inspiration.

My work in Nepal was supported by a series of generous grants. I would like to thank the United States Environmental Protection Agency STAR Fellowship Program, the Yale Program in Agrarian Studies, the Yale Center for International and Area Studies, the Tropical Resources Institute at Yale, and the Edna Bailey Sussman Fund for enabling travel and research in Nepal for the duration of the preliminary field research on which this book is based. I am grateful to the New York University Research Challenge Fund and the NYU Program in Metropolitan Studies for assistance in periods of follow-up research, and to the Hong Kong Institute for the Humanities and Social Sciences for support that enables continued scholarly conversations and collaborations that build outward from this work. Long before my interest in Nepal led me to pursue graduate study, support from the Thomas J. Watson Foundation made possible a year of residence and independent research in Kathmandu.

My bookmaking journey benefited from the detailed attention and expertise of a generous set of anonymous reviewers, the supportive editorial assistance of Valerie Millholland and Miriam Angress, and the creative input of Katie Osbourne. Mara Naselli devoted her time and masterful eye to the formative early stages of crafting the manuscript. Ashwini Srinivasmohan provided invaluable assistance in the book's final editorial stages. I am grateful to each for the positive and lasting impact they have left on my thinking, my writing, and on this book. The shortcomings that remain are my own.

My greatest debts by far are to my family. From my earliest memories, my parents, Nancy Holzrichter Rademacher and Ronald Rademacher, encouraged me to think critically and to engage my curiosities. Their support, in all senses of the term, made this book not only possible but thinkable. I am also grateful beyond measure to Steven Curtis, who was an unfailing creative and intellectual companion during the many years of fieldwork recounted in this book.

To complete this book is, above all, a great privilege. It is my sincere hope that it honors those whose generous gifts have brought it into being. In the end, I remain a student of urban ecologies, and my hope for this volume is that it opens their complex and urgent questions more fully.

Sections of chapters 5, 6, and 7 include data or analytical points previously presented in the following published material: "Restoration and Revival: Remembering the Bagmati Civilization," in *Symbolic Ecologies: Nature and Society in the Himalaya*, ed. Arjun Guneratne (New York: Routledge, 2010); "When Is Housing an Environmental Problem? Reforming Informality in Kathmandu," *Current Anthropology* 50(4) (2009), 513–34; "Marking Remembrance: Nation and Ecology in Two Riverbank Pillars in Kathmandu," in *Narrating the Nation in Public Spaces: Memory, Race, and Empire*, ed. Daniel Walkowitz and

Lisa Maya Knauer (Durham: Duke University Press, 2009); "Fluid City, Solid State: Urban Environmental Territory in a State of Emergency," *City and Society* 20(1) (2008); "Farewell to the Bagmati Civilization: Losing Riverscape and Nation in Kathmandu," *National Identities* 9(1) (2007); "A 'Chaos' Ecology: Democratization and Urban Environmental Decline in Kathmandu," in *Contentious Politics and Democratization in Nepal: The Maoist Insurgency, Identity Politics, and Social Mobilization since 1990*, ed. Mahendra Lawoti (Delhi: Sage Publications, 2007).

Map 1. A regional map of Nepal. *(Created by Katie Osbourne)*

Introduction
A Riverscape Undone

A MARCH

The bright, festive invitation contained an unexpectedly somber plea: "Help save Bagmati," it read, listing a roster of prominent figures who would attend a late-December rally to raise awareness about the plight of Kathmandu's degraded rivers. It was 2002, and Kathmandu's residents had witnessed a rapid decline in river conditions for more than a decade. Certain residents, activists, and policymakers had decided they had seen enough, and the next step was to make a collective and visible demand for change. A well-known Nepali environmental group organized the event, inviting officials, activists, journalists, and tourists to the two-day program of riverbank walks. Each walk would terminate at Teku, the confluence of the Bagmati and Bishnumati rivers. There, from a large stage, river improvement advocates would deliver passionate speeches about the rivers' past, present, and future. Each would argue that, without action, the rivers and the city would suffer irreversible losses.

The rally began with a parade-like procession. A Newar *dāphā* ensemble marched at the head, its musicians clad in black vests and hats (*topī*s) and sounding cymbals, *mādal*s, and other large drums. NGO (nongovernmental) workers, journalists, and invited guests followed, some sporting large VIP tags pinned to shimmering blue ribbons. At the rear of the group marched a handful of foreign tourists, dressed in the colorful patterns sold in the shops of Kathmandu's tourist district. Each held aloft a sign neatly written in Nepali, furnished by the event organizers.

The combination of musicians, dignitaries, and activists gave the procession the simultaneous air of cultural celebration and political protest. But the setting, river channels choked with garbage and sewage and riverbanks host to seemingly countless shacks of the city's poor, framed the festival as a portrait of despair. As the marchers' bright banners bobbed along the riverbank, the main speaker, the cultural heritage activist Huta Ram Baidya, stopped periodically to map precisely where, how, and why the rivers suffered what many considered both ecological and cultural abuse.

Public demonstrations of any kind were at this time still relatively new in Nepal's capital. Open, public assembly had been possible again only since the People's Movement reintroduced this freedom in 1990. The procession to raise river awareness was thus a democratic practice in a new democratic period only a decade old—a period that had also witnessed extreme environmental deterioration. For many in Kathmandu, the polluted Bagmati River had come to signify democracy's general failure to live up to its promise. It was also a symbolic marker of various forms of cultural, religious, and national disorder. For those who marched in the procession and for many others in the capital at large, river degradation suggested deep and interconnected environmental, social, and political dysfunction.

Marchers huddled around Baidya as he spoke over the percussive sounds of the dāphā; despite his megaphone, the elderly speaker was difficult to hear. He stopped to address the crowd at the site of the UN Park, proposed nearly a decade before but still largely unbuilt. In addition to establishing a recreational riverbank park, the project was intended to commemorate Nepal's contributions as a member state to the United Nations, particularly the Nepalese army, police, and civilians who served in UN peacekeeping missions. In its early stages, park boundaries fluctuated, but by 1999 a master plan covered river corridors along much of the Bagmati and Bishnumati in Kathmandu. Yet the proposed park still existed only in plans and proposals, just one among many unrealized hopes for urban environmental improvement.

The march continued under the Bagmati Bridge and in front of Dashnami Akhada, a temple.[1] I noticed a dignified man in a *daurā suruwāl*, a suit suggesting official government status. His VIP nametag was much larger and brighter than those worn by other honored guests, and I quickly recognized him as a scholar and activist who had recently been named as the minister of water resources. With another foreign woman, I was gently pulled aside by one of the minister's aides. Saying he had something to show us, the aide led us away from the procession and over to the river's edge, where we found the minister waiting. Standing just beneath the Bagmati Bridge, he pointed to the structure's foundation and addressed us in English: "You know what that is, don't you?" he asked. Without waiting for a reply, he continued, "It's a fish ladder!" Indeed, when the new Bagmati Bridge was completed in 1997, a fish ladder was installed in the cement platform lining the bridge's base. As was the case throughout this reach of the river, though, the fish ladder was clogged with plastic bags and household waste, and the primary flow of river water, most likely devoid of fish of any kind, was easily ten feet away. The minister explained that the ironic and sadly unnecessary structure was symptomatic of river management and development gone wrong. "It was built because that's what they were supposed to build," he explained. "Never mind that there are few if any fish species that can survive in such a polluted river." I asked who had built it, but the minister replied that he didn't know, that it was part of an aid project of unknown provenance; the point was *how* it was done. With that, we were left staring at the fish ladder, reminded of the critical role that international development projects like these played in the history of river degradation. The minister, meanwhile, hurried off to rejoin the procession.

By now the march had moved ahead to Kalmochan Temple, where a crowd again huddled around Baidya and his megaphone. The group fixed its gaze toward an *arghajal*, a carved stone platform used in Hindu ritual to lay a dying person in a way that allowed his or her feet to touch the river's edge. This particular structure, however, was crumbling into the dry and eroding riverbank, an unlikely host to such a sacred purpose. Baidya somberly told the marchers that this was the arghajal where his family members had their last rites performed and where, he told his audience, he, too, wanted to die. "I want my feet in *this* water," he said, as if to amplify the very absence of water from this part of the riverbank. His voice grew more forceful and impassioned as he continued: "Should we bring the water *back* to places like this, or should we build parks that assume the water will stay where it is — far, far away from our holy places?" Baidya was referring to designs for the UN Park, plans developed according to the rivers' present, degraded morphology. Any new park construction, Baidya argued, should be undertaken in tandem with morphological restoration that would widen river flow and bring water back to his ancestral arghajal. "We have all seen development," he shouted, "which has given us some things. We talk about rights. But what about our cultural rights? Development has destroyed my cultural rights! . . . Our culture has been broken!" The arghajal, too, was broken—cracked and crumbling from the sunken, caved-in riverbank beneath it.

Solemnity fell over the group as the march continued. The musicians stopped playing, and their music was replaced with the sounds of everyday life along the river. We came to the *ghats*, stone stairways and meeting places that had once led to the river but now provided other uses. Chandra Ghat had become home to a police barrack and Juddha Ghat was now a school. Instead of flowing water below these historically sacred landing places, the riverbed zone now yielded crops—meticulous plantings of maize, vegetables, and banana trees. The river itself flowed so far from where we stood, in fact, that its narrow course was almost impossible to discern in the distance. Toward Juddha Ghat, cultivated riverbank land led to a playground abuzz with shrieks of laughter and games of all kinds. Our riverbank march seemed to have lost its river completely, tracing instead everyday lives folded in the social layers of a changing capital city.

Clothing of various sizes and colors hung out to dry on the fences, bushes, and stone steps of the ghats. Vendors sold tea, biscuits, and cigarettes, and the air swirled with the smell of kerosene cooking fires. Children scrambled around us as their caretakers stretched out on straw *sukul* mats, warming themselves in the winter sun. Some women washed dishes from the morning meal; others sat and stared at the spectacle of our procession. Behind me I heard pieces of marchers' conversations. "*Bāhirabāta āeko mānisharu* [These people are from outside]," one said, referring to those living along the riverbanks in large clusters of informal settlements. I wondered what kind of conversations I would hear if I were positioned instead in the settlements through which we marched, among the migrants who took shelter here and regarded it as home.

Just beyond Juddha Ghat, the procession approached Bansighat, an area long settled by landless migrants (*sukumbāsī*, hereafter, sukumbasi).[2] The territory was unmistakable: a huge cluster of brick, mud, and bamboo dwellings, with walls of thatch covered in cobalt blue plastic tarps, entirely blocked our distant view of the river. Only through an occasional alleyway could we see the narrow river channel in the distance; otherwise, we were in the midst of a bustling neighborhood.

Bansighat is one of the Bagmati's oldest sukumbasi settlements, dating back at least to the 1970s. As the procession moved toward a widened section of road in the middle of the neighborhood, Baidya stopped to address the marchers. They gathered around him, and soon dozens of women, children, and men from the Bansighat settlement joined as well. I felt a sudden uneasiness about what Baidya might say. I knew from our private conversations that he considered riparian settlers to be encroachers whose presence exacerbated river degradation and prevented what he considered to be proper river restoration activities. I feared that a confrontation between Bansighat residents and the marchers might follow if Baidya charged settlers with riverbank encroachment or harm. I drew a deep breath as Baidya stepped forward under a large, old tree. The crowd of residents and marchers fell silent, and he began to speak.

Instead of discussing the riverbank migrant settlement, Baidya focused on the dirt road that we had all just crossed. The very existence of this road, not to mention its proximity to the watercourse, he announced, demonstrated that government officials and planners lacked "cultural respect" for the Bagmati River. Built in 1955, the year that King Mahendra ascended the throne, the road marked the beginning of a long period of struggle between a government that was organized around a strong central authority and periodic experiments with "democratic" diffusions of political power.

Sensitivity to the cultural value of the Bagmati, Baidya continued, would surely have prohibited the kind of modern, developmentalist construction that this riverside road exemplified.[3] "Roads should be built," Baidya shouted forcefully, "at least thirty meters from a water source!" He then pointed across the way to a huge, five-storied home — not a migrant hut but a "legal" mansion of pink cement. "Do we control this," he asked, "or do we let people do whatever they want?" Of the sukumbasi settlement to his back, he said simply that it was another topic, once he did not want to raise. With this, a potential standoff between the marchers and those living informally on the riverbanks was defused, but territorial claims were nevertheless clear. Baidya's rhetorical question "Do we control this?" applied to both settlements of the landless poor and the large homes of elites: Both types of structures encroached on the former riverbed.

By now the early morning chill had given way to an intense noonday sun. The procession continued to a large festival ground near Teku. In preparation for the event, NGO workers had cleared the area of the mounds of trash usually dumped there, and the river flowed at a distance, which diffused its otherwise sharp odor of sewage. An imposing stage adorned with red banners and flowers completed a temporary courtyard made up, on three sides, of booths hosting exhibitions, literature, and displays by various environmental NGOs. A long program of speeches from the stage continued well into the afternoon. By three o'clock, tired and sun-weary, I decided to return home, retracing my steps from the procession earlier that morning.

As I approached Bansighat, some women I didn't recognize stopped me. They'd seen me as part of the march earlier that morning, they said. "Is it true that they're going to evict us?" one woman asked, clutching my sleeve. "What's the news about eviction back there?" The women were visibly frightened, and as we spoke, they explained that earlier that day, after the procession passed, the police followed and announced that the marchers were arranging to have them all forced out. Sukumbasi, they were told, were the true target of the demonstration. One of the women launched desperately into an explanation of her situation. She had lived on the riverbank for eleven years, she said; she and her family were extremely poor and had nowhere else to go.

At that moment, the river conflicts that I had been studying for years replayed in my silent deliberations over how to respond. I heard the same voices of suspicion and caution that environmental activists had been sounding about "encroachment," while the fear this woman expressed and the details of her story invoked the mission of housing activists with whom I had also aligned and worked closely. I was relatively safe for these women to approach, as I was neither Nepali nor fully foreign, and I could speak to and understand them in Nepali. However, I was also potentially dangerous, as one who had come from a rally that the women believed was intended to displace them. Perhaps I was even connected to some of the "development" powers that made decisions about the rivers and, in turn, their very fate. This moment underlined the complexity of river restoration politics, and the place of any researcher attempting to understand them. There was no neutral answer to the women's questions, just as there was no apolitical or disengaged posture from which to study or advocate for river improvement.

I told the women what I had heard in the speeches at Teku. In some presentations, riparian migrant settlements were named as obstacles to river restoration, but few speakers made overt calls for evictions or resettlement. As far as I had understood, the march and festival were intended to raise awareness about the degraded condition of the rivers but were not explicitly about settlements on the riverbanks. I added that I knew that some river activists were hostile toward sukumbasi and all others whom they considered to be out of place on the riverbanks.

My answer was not reassuring. The women continued to detail their personal plights, expressing outrage over the suggestion that their presence fouled the environment any more than the "legal" city residents whose garbage and untreated sewage poured into the rivers daily. I didn't challenge them, and somehow despite my association with the marchers, the women did not challenge me. Instead, after nearly an hour of listening, they asked me to carry their stories back to the people at the festival. "They will listen to you," one woman told me.

The next morning, I walked toward Teku to hear more speeches, tracing the same route from the day before. I met two women I recognized, and one asked immediately if I had seen the newspaper. She waved a copy of *Kāntipur*, pointing to an article about the previous day's march.[4] Deep in the text was a sentence that implied that the municipality planned to evict those in the riparian zone whose dwellings were deemed illegal. I abandoned my plan to return to the festival ground and instead accepted the women's invitation to talk further and drink tea together. One woman's words, wrought with anger and fear, echoed Huta Ram Baidya's own from the day before. "Is this development?" she asked, defiantly. "In development I have lost everything. Our lives," she told me, are just "*sarkhārko khel* [the government's game]."

The contest over who should determine the ideal form and function of the Bagmati River in Kathmandu—that is, who should define river restoration and its contours—took place in a period when ideas of contemporary democratic process and practice were contested, and in some ways nascent, in Nepal. For many, the transition from a monarchy-centered government to a parliamentary democracy after 1990 carried with it great promise for positive and rapid change. But the actual experience during that period was punctuated by disappointment, deterioration, and, as the marchers sought to amplify, intensifying urban ecological decline.

The march to save the Bagmati was at once a reaffirmation that democratic processes could promote certain forms of change and a reminder that, for many, the expectations that had accompanied democracy remained staunchly out of reach. For riparian settlers, a sense that democratic governance was somehow operating through them, and on their behalf, was not only elusive but often simply absurd. Marchers and settlers agreed that environmental conditions on the river were unacceptable, but the precise contours of river degradation — whether they involved housing rights, ritual access, or ecosystem integrity — were urgently contested. The problem of how to define, and reverse, river decline pointed to an equally disputed question in the capital city: Who should be empowered to effect change in Nepal? Could the urban future improve? If so, how?

A PREDICTION

The organization of Kathmandu Valley through the Hindu monarchy has a long history; cultural associations between kingship and divinity there have been traced to the fourteenth century. The kings of Nepal's Shah Dynasty ruled the country from the mid-eighteenth century, when the Valley was conquered — or the country was unified, depending on one's political position.

The Shah Dynasty introduced an uneven but continuous system of cultural strategies to reproduce monarchical rule. These strategies, which persisted through the twentieth century, included royal land grants, networks of patronage, and the development of a state bureaucratic apparatus that favored high-caste Hindus and the ethnic Newar of the Kathmandu Valley. They also involved the promotion of ideas of divine kingship.

The Nepali state has long consolidated its power through formal associations with religious symbolism and practice, and this continued after the 1990 transition to a democratic model of governance. The state retained its constitutional status as a "Hindu Kingdom," and prescriptions against cow slaughter, religious proselytizing, and official observances of Hindu holidays continued in practice even as objections were more freely voiced.[5]

A legacy of monarchical patronage, favoritism, and cultivated moral authority helped to shape the ways that many Nepalis, across caste, class, and ethnicity, imagined change, including post-1990 democratic change. Although nominally democratic aspirations emphasized liberal conceptions of citizenship, some Nepalis retained an enduring belief that ultimate power and responsibility to forge lasting change rested with the monarchy.

One summer day, two years before the march to save the Bagmati, my friend and research assistant, Sama, invited me to accompany her to Pashupatinath. We had visited this temple along the Bagmati many times before, but on this day we went for a particular ritual called *puja*. Sama was an upper-caste Hindu woman whose middle-class family, like the Shah Dynasty itself, originated from the central Himalayan district of Gorkha. She had married, had children, and had grown more and more meticulous in her attention to religious practice since the time we first met in Kathmandu in 1992.

Pashupatinath has long been an important pilgrimage site for orthodox Hindus. Since it houses the only *Jyotir Lingam* in the country, one of a circuit of twelve Shiva temples (the other eleven are in India), it is often called the most sacred Hindu site in Nepal.[6] It is also the historical symbolic center of the Hindu Kingdom of Nepal, for which Lord Pashupatinath was the patron deity. Even before the Shah conquest and establishment of Nepal's Shah Dynasty, Valley royalty supported the temple. Its present structure dates to national unification in 1769, and subsequent Shah kings continued patronage practices after that date. The temple complex sits on the Bagmati riverbank, at a site believed to be in ritual use since the third century BCE.

Pashupatinath bustled with activity on that hot and dusty summer day. Just beyond the entryway to the temple grounds, vendors packed the roadside, marketing their wares. Cones of bright yellow and red powder towered over fruits, flower garlands, butter lamps, and strands of beads; print images of deities large and small gazed from carefully arranged displays. Sama and I made our way inside the compound, gently swept along by a crowd of women carrying trays laden with offerings. Once inside, we stopped together, silently absorbing the scene from a footbridge over the Bagmati that connects two parts of the temple grounds. In view was a main temple building, several smaller structures, and, slightly to the south, the Arya cremation ghats. Below us lay the largely dry Bagmati riverbed, its banks scattered with mounds of discarded plastic bags and garbage. Small leaf plates carrying ritual offerings dotted the sandy soil, and the scents of incense and burning oil lamps drifted over a pungent odor of sewage.

Sama shook her head. Of all the places along the city's polluted rivers, she said softly, as if speaking only to herself, Pashupatinath affected her most deeply. "When I see the river *here*, the sadness feels very deep. What has to change to make the Bagmati clean again?" She then turned to the cremation ghats and reminded me of their purpose. "Royalty," she said firmly.

"When the king dies," Sama explained, "his son must attend the funeral rites. On that day, he will *have* to come here and see the Bagmati River in this terrible state. He has no choice. The cremation has to take place here." After that, she said, it would be impossible for the royal family to ignore the river, and surely once they took notice, the monarch himself would ensure that the Bagmati's ecological and cultural integrity was restored. "When King Birendra dies — only then will the river change."

Along with a largely elite subset of Kathmanduites, Sama believed that, despite what at that time had been a decade of democracy, only the royal family retained the ultimate responsibility for, and power to change, the Nepali nation, its people, and its landscape. Her view suggested the lingering traction of a modern Nepali nationalist imaginary, forged during the Panchayat era, in which the king was promoted as the unifier of the vast and complex assemblage of caste, religious, ethnic, and linguistic groups that together form the Nepali nation-state. That king was the ultimate purveyor of order and the symbolic center of a national idea, a focal point of unity and belonging among an otherwise extremely heterogeneous populace. For those who shared Sama's view, this sense of enduring royal power and responsibility extended quite logically to the capital city's ecology. As King Mahendra had ensured in the Constitution of 1969, and his successor King Birendra reinforced in 1991, the king held definitive authority in times of crisis; Sama anchored her hope for a solution to the Bagmati's ecological problems on this power. It was ultimately the king who could provide Nepal with the moral and practical guidance it would need to reverse the seemingly intractable condition of the city and its riverscape, she reasoned.

Less than a year later, following events that neither of us could have imagined from the footbridge at Pashupatinath, King Birendra's cremation actually came to pass. With it, also in ways previously unimaginable, the logic of the nation, the place of the capital, and the future of the Bagmati River and its meanings were profoundly changed. But the content of the cremation rituals, and the form of the changes that they signaled, confounded anything we might have envisioned that day in 2000.

Dipendra, King Birendra's son, did not attend the funeral along the ailing Bagmati River at Arya Ghat. Instead, as the king and queen were being cremated, Dipendra lay unconscious in a military hospital, having reportedly killed his father, mother, and other members of his family, and then attempted to kill himself. Instead of the king's son, his brother, Gyanendra, stood before the Bagmati at Pashupatinath, even as many in the capital staged angry demonstrations to voice their suspicion that Gyanendra had himself somehow orchestrated the murders. The death of King Birendra in a horrific mass killing disrupted any anticipation of a smooth transfer of power from the king to his son, and the aftermath of the royal cremation at Pashupatinath inspired not the reversal of a river's ecological decline but the acceleration of a profound and still unfolding political transformation.

Meanwhile, largely outside the capital, a mass movement for revolutionary change proceeded with great speed and effect, violently and formidably challenging the fundamental form of the polity itself. Nepali government forces and fighters for the Communist Party Nepal-Maoist (CPN-M) battled for control of the country's rural districts in a conflict that came to be known as the People's War (*jan yuddha*). Kathmandu remained largely insulated from the brutal violence, but as the conflict waged on, Valley residents feared that its spread to the city was imminent. A host of unforeseen changes were expected in its wake.

SHAPING URBAN NATURE IN AN UNSETTLED STATE

In the march to save the Bagmati and in Sama's hope for change after the king's death, actors engaged questions of power and political transformation through ideas of environmental change. The actors described above all linked their aspirations for order in Nepal's capital to perceptions and practices related to their urban environment: Baidya criticized modern government and development for having "stolen [his] cultural rights"; a marginalized settler saw her territorial survival as a "game" of the government; and Sama was committed to an ultimate faith that a monarch would enact river revival. Sama's river restoration logic was refracted through a belief in the enduring legitimacy of a monarchical state, while others sought the powers and rights suggested by ideas of democracy. For Baidya, and perhaps the migrant woman who shared her exasperation, democratic "rights" were yet to be fully realized; for Sama, the new king needed only the appropriate time and ritual setting to bring about necessary change.

In this book I attempt, through ethnographic experience, to elaborate and better understand relationships between political and environmental transformation in a particular time and place. I focus on the degraded urban reaches of the Bagmati and Bishnumati rivers, which converge in the heart of Nepal's capital city, and on the multiple networks of activism and bureaucratic practice that engaged with the rivers, and extended outward from them.

Through the 1990s and into the twenty-first century, severe ecological degradation captured the attention of elites, activists, and development professionals in the capital. Anxieties over the condition of the rivers, and in turn the city itself, punctuated broader debates about Nepal's urban future. I will explore these anxieties and the contests over meaning and territory that arose from river rehabilitation efforts. Through a study of ideologies and practices of reviving the rivers, I explore how urban nature was experienced, in part, through claims about cultural meaning, history, and territorial belonging. By considering ecology in terms of social practice, and describing social practice through ethnographic experience, I approach the Bagmati and Bishnumati case as a way to study how urban nature and urban social life are mutually produced, reinforced, and, ultimately, changed.

I will situate a set of organized urban river restoration efforts in political context by drawing on my fieldwork between 1997 and 2008. This was an extremely volatile period in Nepal, during which the ruling government was in the throes of rapid and continuous transition. From the turbulent and incomplete conversion to a democratic constitutional monarchy in 1990–91 up until the "People's War" between the Royal Nepal Army (RNA) and the army of the CPN-M, the stability and sovereignty of the state remained in flux and were constantly contested. The violent conflict between state authorities (the Nepal police, and, later, the RNA) and the CPN-M army began in 1996. The resulting conflict claimed over 13,000 lives by 2006. Following the massacre of nearly all of Nepal's royal family in June 2001, the new king, Gyanendra, consolidated authoritarian rule, suspending what remained of nascent democratic processes. This exacerbated state instability and further polarized the positions of those holding political power and those seeking radical change.[7]

The latter period of the People's War was marked by a national state of emergency, declared after Maoist fighters directly attacked the RNA for the first time (previous violence was between the Nepal Police and the Maoist army). In November 2001, after peace negotiators failed to reach an agreement in a third round of talks between His Majesty's Government and the CPN-M, the Maoist army launched attacks in Surkhet, Dang, Syangja, and Salleri. At least fourteen RNA soldiers, fifty police officers, and several other government officials were killed; the number of Maoist casualties was unclear. Although as many as sixty Maoists were reported killed, only fifteen bodies were officially acknowledged as being recovered. After Prime Minister Sher Bahadur Deuba announced an emergency on November 26, the government mobilized the RNA to fight Maoist insurgents, whom Deuba officially declared to be "terrorists." The cabinet enacted a Terrorist and Disruptive Activities Ordinance, authorizing arrests without due process and facilitating new controls over media and information. These controls extended to public life and free expression, rendering public discourse on topics like urban social and environmental policy nearly mute.

On February 1, 2005, King Gyanendra seized direct power, which lasted until massive protests around the country led the king to reinstate Parliament in April 2006. Through a declaration, the reinstated Parliament officially stripped the monarchy of its power in May 2006, and elections for a Constituent Assembly, expected to draft a new democratic constitution for Nepal, took place in April 2008.

It is in this biophysical and political context that I conducted field research among activists and officials involved in river improvement debates and interventions on the Bagmati and Bishnumati. During twenty cumulative months of fieldwork over multiple periods between 1997 and 2003 in Kathmandu, I did extensive participant observation and I conducted periodic semistructured interviews with those in activist and development organizations actively engaged in the planning, execution, or contestation of the Bagmati and Bishnumati improvement initiatives. In the summers of 2006 and 2008, I returned to Kathmandu for follow-up participant observation and semistructured interviews with informants from the 1997–2003 study.

Taking The Bagmati Basin Management Strategy (Stanley International et al. 1994) and the Kathmandu Urban Development Project Bishnumati Corridor Environmental Improvement Program (His Majesty's Government and Asian Development Bank [HMG/ADB] 1991) as starting points, I traced river improvement actions and reactions in an effort to follow the conflicts (Marcus 1995) that animated river restoration politics. My interlocutors included bureaucrats and development professionals in charge of specific interventions for river improvement, including those involved with the Bishnumati Corridor Environmental Improvement Program and the UN Park Project, as well as NGO workers and activists for housing rights and river cleanup, including the shelter advocacy organization Lumanti, the culturally focused Save the Bagmati Campaign, and the largely urban, expatriate Friends of Bagmati. These individuals and groups were embedded in networks through which critical representations of, and knowledge about, Kathmandu's environmental situation were generated, disseminated, and contested.

I followed three main groups as they tried to assert control over the restoration of Kathmandu's riverscape: (1) state and development experts working for river restoration, (2) cultural heritage activists concerned with restoring the templescape and river-centered religious practice, and (3) housing advocates for the tens of thousands of poor migrants who settled along the riverbanks and in exposed riverbeds over the decades immediately preceding the period of the study. Each group made competing assertions about the cultural meanings and histories that the rivers embodied, and each claimed entitlement to riverscape territory. Yet all shared a common frustration with the pace and tenacity of river degradation and with the serial failure of official river restoration initiatives.

In this transformative period, river restoration was a lens through which to observe intersections between urban ecology and state making.[8] Since efforts to ensure, create, or imagine ecological stability regularly intersected with aspirations for political stability, river improvement initiatives and conflicts demonstrated important dynamics in the reproduction or contestation of cultural ideas of the nation-state.[9] These conflicts also involved the construction of new affinities understood to foster social cohesion where other ways of marking sameness and difference could not or did not. Such environmental affinities derived from complex combinations of arguments about indigeneity, competing ideas about governance, and moral logics through which Nepal's urban future could be organized and ordered.

A single ethnography cannot capture the vast complexity and range of Kathmandu's political and environmental transformation in recent years. I attempt, instead, to focus and reflect on a subset of actors and projects. By doing this, I offer an analysis of the ways that ecological ideas and practices conditioned certain ideas about universality through which political forms and social marginalities were justified, reinforced, or produced. By considering environmental and political change as being mutually produced, I emphasize how modern ecology formed discursive and practical terrain for both a state seeking to retain its legitimacy and for those who sought to challenge or reshape that very state.

In order to focus on social ideas and practices of urban ecology, I offer only a partial accounting of the wealth of biophysical data through which riverscape change and management has been captured, catalogued, and operationalized since 1950. My approach proceeds from the persistence of river degradation, and indeed its intensification, even as the scientific management of river degradation was well underway. In certain ways, this persistence pointed away from the content of biophysical data and reports and toward the complex of social interactions through which river degradation was framed, debated, and made meaningful as an urban ecological problem.

Addressing urban ecology in this manner is not intended to suggest that the rivers were, or that the environment ever is, simply and exclusively a social construct. Scientific characterizations of ecosystem change are as important as the social processes through which they are produced and within which they are encountered, interpreted, and disseminated. Biophysical settings, including dense urban landscapes, are not infinitely malleable, regardless of our recognition that social forces are crucial for delineating the form and content of environmental categories (Benton 1989; Gupta and Ferguson 1992; Mosse 1997).

Any concept of nature is, in fact, "produced through the interaction of biophysical processes that have a life of their own and human disturbance of the biophysical" (Sivaramakrishnan 1992:282). Thus, nature is conceived out of this interaction between the human and the biophysical. Sivaramakrishnan continues, "Human agency in the environment, mediated by social institutions, may flow from cultural representations of processes in 'nature' but we cannot forget the ways in which representations are formed in lived experience of social relations and environmental change" (ibid.). Understanding how biophysical constraints and social imaginings converge on a given landscape, then, is fundamental to studying ecology as the sets of experience and action (Peet and Watts 1996; Redclift and Benton 1994) that I will call "ecology in practice" within this book.

In this book, I outline the biophysical contours of river degradation (chapter 3), but I focus on the range of social processes through which they were fashioned and engaged as an ecological problem situated in time and place. My intention here is to underline that the riverscape — a facet of urban nature — was produced at a nexus of cultural and biophysical processes, and to illuminate the social dimensions of this nexus. To do this, I pay attention to differently positioned actors involved in river change and to the ways that they defined, understood, and represented that change as problematic. I focus on competing claims about the stakes of river transformation and on assertions about the sociopolitical formations that would best facilitate river restoration. In short, I ask, what did river degradation and restoration *mean* for different people during a specific political moment?

A CITY IN CRISIS

Nepal's capital city is located in the Kathmandu Valley, an area contiguous with the 600-square-kilometer Upper Bagmati Basin, which drains the Bagmati and Bishnumati rivers. The two rivers converge in the heart of the city at Teku Dovan, a temple complex that marks a mythological point of origin of Kathmandu. Lined with Hindu and Buddhist temples that attest to centuries of cultural practice and political change, these rivers hold historical and ecological importance for Kathmandu, the city that has long occupied the symbolic and material center of the Nepali nation-state.

After they converge, the Bagmati and Bishnumati flow through the Middle and Lower Bagmati Basins, then eventually join with the Ganges River. The Bagmati and Bishnumati are unusual in Nepal, for, unlike the large glacial rivers more often associated with the Himalayan region, they are replenished not by snow and ice but by rainfall and springs. Since the head reaches of both rivers are located within the Valley, the amount of surface water and the capacity for the rivers to assimilate wastes are relatively limited.

During the 1990s, as unprecedented urban growth dramatically changed Kathmandu's physical environment, the Bagmati and Bishnumati rivers were highlighted among the urban features characterized in official and popular discourses as being increasingly degraded.[10] River degradation signaled a range of conditions, including severely reduced water flow and quality, significant morphological changes, and, some argued, the loss of cultural and religious values historically attributed to the rivers. Policy and development studies, such as The Bagmati Basin Management Strategy (Stanley International et al. 1994), emphasized four main causes of river deterioration inside the urban area: the discharge of nearly all of the city's sewage—

completely untreated—directly into the rivers; the widespread dumping of garbage into the rivers and on their banks; sand mining in riverbeds and banks, which supplied mortar and cement materials to the city's booming construction industry; and human encroachment on the banks, floodplains, and riverbeds exposed by channelization. Sand mining was especially blamed for significant morphological change and severely channelized flow patterns in both rivers.

Rapid urban growth produced a level of housing demand that overwhelmed existing housing stock, resulting in widespread informal settlement throughout the city. Because morphological change had produced large sand flats in the riparian zone, common destinations for those seeking land were the riverbanks and exposed riverbeds of the Bishnumati and Bagmati.

In 1991, the Asian Development Bank estimated that informal riparian settlements on the riverbanks were growing at 12 percent annually, a rate twice that of the city's population itself (HMG/ADB 1991). By 2001, a significant part of the urban river corridor was lined with semipermanent structures and settlers asserting rights to the land they occupied.

Worsening river conditions motivated a variety of state- and internationally sponsored restoration initiatives that, in turn, drew steady criticism. Some critics charged that official restoration activities were consistently insensitive to, or abusive of, various characterizations of the rivers' cultural significance, particularly Hindu constructions of the rivers' sacred qualities and ritual power.[11] Others pointed to the ways that river restoration projects threatened the security of tens of thousands of landless migrants settled on the exposed riverbed lands of the riparian zone. Few would dispute that the rivers were degraded or that degradation had both scientific and cultural causes. Yet precisely *why* river quality continued to deteriorate and exactly *how* to reverse it were points of ongoing contestation. While the existence and persistence of something called degradation was a matter of overwhelming agreement, its contours, and consequent pathways to actions understood as restoration, were fraught with intense disagreement.

Kathmandu's environment was by no means the only—or indeed even the most prevalent—facet of urban life enveloped in discourses of crisis. Official attempts to improve the urban environment were undertaken against a backdrop of persistent anxieties about the future form and functionality of the polity itself. As the People's War and other political events intensified over successive periods of field research, river-related discussions increasingly overlapped with speculations about the future political configuration of the capital city and the Nepali state. Uncertainty about this future contributed to a broadly defined political crisis widely characterized as unprecedented.

But crisis itself, as a frame for understanding Nepal, is by no means new; crises involving irreparable loss of one form or another—with intersecting components both natural and cultural—have characterized development and academic depictions of Nepal for decades. Titles like *Losing Ground* (Eckholm 1976), *Nepal in Crisis* (Blaikie, Cameron, and Seddon 1980), and "Losing Shangri-La?" (Shrestha 1998) refer to a long modern history of vulnerability and socioeconomic suffering, but they also form an almost continuous narrative of intractability and dysfunctionality in modern, particularly developmentalist representations of Nepal's past and future.

During the 1990s, popular impressions of decline on the urban reaches of the Bagmati and Bishnumati rivers regularly raised specters of irreversible losses of diversity, history, and identity. Consider, for example, the journalist and public intellectual Kanak Mani Dixit's (1995) characterization of Kathmandu Valley's cultural degradation in the wake of development and modernization. This view circulated among the publication's readership of urban elites and expatriate development professionals:

> Even as the physical structure of the city undergoes transformation, so too numerous traditions are dead or dying. As patronage disappears, musical conventions and techniques of making masks, statues and musical instruments, or stonework and brickwork are forgotten. The tradition of beating *nagara* drums in Bhadgaon has had to be artificially resuscitated. Numerous *jatras* are affected because their ritual routes have disappeared under new neighborhoods. The chariot of Machhendranath (Bungdeo) becomes increasingly dangerous to pull through Patan town because the wide berths required for its stay ropes are no longer available. Photographers are frustrated as they try to locate the vantage points from which pictures were taken in the past. Macadam, concrete pylons, brick walls, high tension lines, tele-

> phone wires have conspired to change landscapes. Brick and cement neighborhoods sprang out of rice paddies in Baneswar, Naya Baneswar, Koteswar, Dhobi Khola, Lazimpath, Chakupath. . . . It did not matter that there were no roads, sewers, water or electricity. Certainly, there was no planning. There must have been a better way for the Valley town to develop. The Great Earthquake of 1934 destroyed much of old Kathmandu. What the Valley is seeing now is a continuous and devastating cultural earthquake.[12]

Through these sentiments, Dixit suggested to his audience a need to rethink the cultural costs of development and modernization, and pointed to concerns over the potentially homogenizing power of developmentalist ideas of progress.

Meanwhile, some scholars reflected on Nepal's modern development experience by asking how and why the country continued to experience extreme poverty and relatively stagnant economic growth despite massive inputs of aid from bilateral and multilateral development institutions. Scholars noticed how specific configurations of state governance and development institutions seemed to limit, rather than expand, Nepal's capacity to effect positive social, economic, and environmental change. Consistent with poststructural development analyses, scholars began to understand Nepal's overall experience of modern crisis as, at some levels, being the complex product of development institutions themselves, rather than a set of problems that could be solved through perfectly targeted policy instruments. In a well-documented case of Himalayan deforestation, for instance (Metz 2010), scholars elaborated on the limits of reforestation initiatives or community forestry projects for "solving" the problem of deforestation, while pointing to the complex ways the problem itself was produced (Thompson, Warburton, and Hatley 1986; T. Campbell 1992:92; Guthman 1997).[13] In fact, some argued, very basic attempts to define development problems and solutions often masked the important power asymmetries that contributed to them (Crush 1995; Escobar 1994; Pigg 1996; Ferguson 1994).

With Nepal's long legacy of environmental, political, and economic crises in mind, I was at first hesitant to understand the plight of the capital's rivers as a crisis for which there could be an immediate and comprehensive policy response. Yet crisis was precisely the frame that my informants consistently employed. Toward the end of the 1990s, some of the activists, development workers, and bureaucrats with whom I worked began to convey a sense that both the environmental situation in the capital and the political situation in the country were truly unprecedented and, therefore, urgent as never before. The rather naturalized undercurrent of "crisis as usual" in Nepal's long-standing struggle with maldevelopment and environmental stress was supplanted by an emergent sense that the particular convergence of ecological and political change embodied by, but not confined to, the Bagmati and Bishnumati rivers, was distinct and *un*usual. It became clear that fear of the future of a layered and heterogeneous Nepali state[14] was a crucial aspect of perceptions of river degradation itself. I therefore faced an analytical puzzle: If crisis itself formed a discursive undercurrent for much of Nepal's modern developmentalist history, how might one recognize and understand what informants explicitly described as an *unusual* crisis? Furthermore, how might one imagine the lived experience of state transformation as a fundamental aspect of ecology in practice?

This puzzle attached the Kathmandu case to a range of localities regularly categorized as sites of so-called permanent crisis, in which the turning point suggested by the term is assumed to be chronic and ongoing rather than acute. I had to grapple, then, with the sociocultural meaning of a crisis in which both a biophysical condition and its attendant social body were not only intertwined, but also mutually implicated in a future of irreversible decline.

If it was possible at all, reversing the Bagmati-Bishnumati crisis hinged on reimagining and remaking the polity itself. In this sense, environmentalism and state making were inseparable projects; each was potentially emancipatory in a way that ensured the vitality and possibility of the other. This raised a further critical question: What political ecological order would ensure a more durable future for Nepal's capital city? What socionatural configuration—what ideal combination of spatial form and social practice—would bring about, and reproduce, vitality for an otherwise critically wounded and chronically suffering city? How was power organized in such a configuration?

A CRISIS OF CITIES

Kathmandu's urban environmental crisis nests into a larger, emergent set of development anxieties that frame a great deal of policy literature on twenty-first-century cities. These anxieties are associated not only with urbanization per se but with its pace and its primary geographic location in rapid-growth cities in the southern hemisphere. These cities, broadly categorized as cities of the global South, are understood in transnational discourse as being mired in varying combinations of intractable poverty and environmental disorder and as being linked to a present or expected "urban explosion" with potentially catastrophic socioenvironmental implications.[15] Dawson and Edwards (2004:6) capture the perceived urgency of this predicament when they write, "The megacities of the global South embody the most extreme instances of economic injustice, ecological unsustainability, and spatial apartheid *ever confronted by humanity*" (emphasis added).

In general, Nepal occupies a noticeable place in global narratives of this crisis, particularly in terms of its conditions of extreme poverty and inadequate shelter. UN-HABITAT's (2003) inventory titled *The Challenge of Slums*, for example, lists Nepal fourth—just below Ethiopia, Chad, and Afghanistan, among the countries with the world's highest percentage of slum dwellers.[16] While not a fixture of global megacity (and more recently, hypercity[17]) debates, Kathmandu has long ranked among South Asia's fastest-growing cities. For nearly two decades, Kathmandu has witnessed annual urban growth rates of between 6 and 7 percent, making it the largest urban center in the nation-state, a strategic center of government bureaucracy, and a rapid-growth capital city whose inhabitants navigate a combination of political unrest and environmental stresses.

Contemporary urbanization is often framed as a new and unprecedented historical condition—and crisis—and concerns about it resonate across disciplines and practices associated with cities. Scholarly work calls for a complete rethinking of "the urban" as an object of study (e.g., Amin and Thrift 2002, Low 1999), while policy literature nervously proclaims that for the first time in human history the majority of the world's population now resides in cities (UN Population Division 2003).[18] Against a backdrop of extraordinary wealth and equally extraordinary wealth disparities (Dawson and Edwards 2004, Sen 2002), scholars and policymakers question the socioecological consequences of rapid urbanization with increasing alarm, asking whether, and how, contemporary urban settings might be re-thought and re-invented as more sustainable, "livable cities" (Evans 2002). The past decade has witnessed a shift from largely passive attention to something called "urban ecology" to including urban ecology's natural and social components among our most pressing global concerns.[19]

Special urban issues of scholarly journals, ranging from *Social Text* (2004) to *Science* (2008), supplement a wealth of new literature on the contemporary global urban condition. Recent titles like *Planet of Slums* (M. Davis 2006), *Shadow Cities* (Neuwirth 2006), and *Maximum City* (Mehta 2004) have both constructed and reinforced understandings of a global, and yet simultaneously Southern, urban predicament marked by seemingly intractable poverty, marginality, and uncontrolled growth and environmental degradation.

As is often the case with the narratives through which we seek to forge coherent ideas of "the global" (Tsing 2000, 2005), the physical and conceptual geographies of urbanization's dire environmental consequences are at best uneven. Indeed, those sites collectively marked by the most urgent urban socioecological problems, and commonly categorized as "cities of the global South,"[20] map the conceptual locus of a contemporary and future urban predicament—giving spatiality to the present or expected "urban explosion" and signaling potentially catastrophic socioenvironmental implications.[21]

In discussing the theoretical category of the global South, Vyjayanthi Rao notes how the literature that discusses the area functions as "a shorthand for a certain kind of work that takes an understanding of the South as its point of departure en route to a theory of globalization" (2006:226). She further notes that, insofar as the global South category promotes historicist thinking, or foregrounds empire as the historical condition within which the very idea of the global is considered, such literature marks an important shift in narratives of globalization. The Social Service National Coordination Act was passed in 1977, forming the SSNCC as a service delivery vehicle. The act required the official registration of all social organizations in the country. A cap was placed on the total number of organizations that could register, and registered organizations could not alter their operations in any way without prior consent from the SSNCC. Each year local NGOs risked not having their registration renewed, so maintaining political favor was essential to survival. All foreign aid donors were required to deposit their funds into the council's account, and disbursement was controlled through it. However, to signal the worst of a planetary urban predicament through the category "global South" also tends in practice to fix into place expectations of deeply dysfunctional socio-environmental forms and to background those processes through which "Southern" conditions are reproduced—processes that likely require constant movement across discursive and material binaries like a North/South divide.[22]

It may be the case that, in a manner similar to the way that an overarching aspiration for something called development focused scholarly and policy analytics on the "dysfunctional" polities and economies of the global South in the late twentieth century (cf. Escobar 1994; Ferguson 1994; Crush 1995; Greenough and Tsing 2003), environmental improvement, and in particular a condition called sustainability, anchors our attention to, and in turn reproduces, a "global South" in the early twenty-first century. Therefore, a particularly Southern urban predicament, marked by expectations and experiences of tenacious forms of urban disorder, suffering, and disaster, is embedded within the global predicament of unprecedented urbanization. Much of this problem is represented as being ecological.

A prevalent facet of the Southern urban predicament, for instance, is the proliferation of informal housing. Resurgent attention to the public health stresses and environmental risks associated with slum, squatter, and otherwise "informal" settlements has brought large-scale incidences of informal housing directly into the purview of urban ecological sustainability (cf. Hardoy and Satterthwaite 1989; Emmel and Soussan 2001; M. Davis 2006) and of urban environmental management (cf. Main and Williams 1994; Evans 2002).[23] At the same time, the aforementioned "slum" and the extreme forms of exclusion it symbolizes assume a place in particular spatial renderings of the history of globalization (cf. Das 2003; Rao 2006; Gandy 2006). This is consistent with the case of Kathmandu, where the question of sukumbasi settlements was an almost automatic feature of conventional diagnoses of degradation and restoration of the Bagmati and Bishnumati riverscapes (chapter 6).

In some ways, the surge in attention to certain aspects of global urbanization continues longstanding engagements with, and concerns about, the future of cities, modernization, and social life (AlSayyad 2003:8). Early schools of anthropology, including the Manchester School (Werbner 1984), and work that motivated the emergence of urban planning itself (Hall 1988) took interest in the relationship between urban environments and social, political, and cultural life. Mid- and late twentieth-century scholarship on urbanization refined the "urban question" by investigating the relationship between urban processes and transnational capitalism (Castells 1977; Harvey 1990; Sassen 1991), and eventually popularized the idea of the "global" (Abu-Lughod 1999) or "world" (Hannerz 1996) city, the "global circuits" (Sassen 2002) on which such a city could be mapped, and the power of cities as organizing centers of capital and politics (Sassen 1991). Attendant questions of scale and state power (Brenner 1998), the future relevance of nation-states (Chatterjee 1986), the emergence of new forms of citizenship (Holston and Appadurai 1999), and questions about the future of social movements (Mayer 2000) all elaborated on parallel processes of globalization and urbanization.

The idea that both the urban form and its attendant socialities strain the integrity, function, and even existence, of nature, ecology, and the environment has its own long and complex genealogy (e.g., Hall 1988:86–135). When combined with modern fears about cities as centers of rapid population growth, this idea assumes a particularly ecological urgency. Ever since *Population Bomb* (Ehrlich 1968) conjured neo-Malthusian specters within Western environmentalism, images of humanity's growth exceeding the ecological limits of its planetary habitat have maintained a tenacious hold on environmentalist politics and have catalyzed public and policy alarm. Despite far-reaching efforts by both academics and activists to refocus population debates into discussions of relative consumption patterns, images of uncontrolled growth of the world's human population—made even more spectral by the relegation of this population to spaces of "permanent marginality" (Gandy 2006)—retain significant rhetorical and material power.[24]

But the environmental crisis of cities, and the environmental crisis of Kathmandu, must also be understood in terms of struggles over power, knowledge, and governance. The officials and agencies that frame environmental conservation problems, and that shape measures to solve them, produce the environment through their expert, authoritative knowledge, and in doing so they encounter, and sometimes supplant, other forms of knowledge and ways of knowing, shaping, and intervening in the environment (T. Campbell 1992:84). Understanding an urban environmental problem, then, particularly one that has attained "crisis" status, requires that we engage the social relations of power that produce, disseminate, and motivate action around that problem in a particular place. The urban global South location of the environmental crisis in Kathmandu matters, because the local aspects of river degradation were always refracted through global perceptions that Southern cities pose the most pernicious environmental problems.

I, then, sketch the contours of ecological practices and diagnostics in the contemporary context of crisis—in a capital city, its national context, and its hemispheric location. The study site—a polluted river—extends to each of these at times, and its meanings and function are layered and co-created by those who engage with it. The multiscaled social complexion of a singular biophysical entity—a river—undermines the presumption of a single analytic through which to define and diagnose "environmental improvement." Activists and river restoration groups are the central figures in this analysis of river decline, because they defined the meaning of biophysical change, the parameters of river crisis, and the interventions that might operationalize appropriate solutions.

THE SOCIAL LIFE OF URBAN ECOLOGY

Assembling biophysical accounts of nature, whether in the city, suburb, or countryside, is conventionally the domain of natural scientists, engineers, and planners—those knowledge producers whose work most directly translates to prescriptions for management and intervention (Mitchell 2002:30). Their modes of inquiry and the languages through which they convey their diagnostics tend to occupy a privileged place among the many ways of knowing and experiencing the socionatural environment.[25] Yet while scientists, designers, and policymakers may produce authoritative accounts of environmental problems and solutions in powerful domains of action, their methods and analyses often yield an incomplete understanding of environmental problems. In particular, they often fail to understand why people undertake specific forms of action in relation to their environment. We are often left to seek other analytical strategies if our objective is to discern the meanings that animate and motivate people's interactions with nature.

Yet science remains an essential and fundamental form of knowledge for understanding environmental processes and change. In the ecosystem sciences, urban ecology marks a specific subdiscipline wherein cities are theorized and modeled as ecosystems, host to specific combinations of stresses, disturbances, and structures that affect nutrient cycling, hydrological processes, air and soil quality, vegetative cover, and a range of other parameters (McDonnell and Pickett 1990; Pickett 1997; Walbridge 1998). By engaging with urban contexts, researchers of urban ecosystem ecology have also highlighted some important limitations of a conventional ecology that regards cities as antithetical to natural spaces and processes. This has in many ways driven a rethinking—albeit confined to particular epistemological parameters—of human-nature interactions and the extent to which "nature" must be scientifically understood as being fundamentally engaged with human activity (e.g., Rebele 1994; Pickett et al. 2001; McKinney 2002; Pickett and Cadenasso 2002; Grimm, Baker, and Hope 2003). The result is a set of scientific models that explicitly link human and natural components of singularly conceived urban ecosystems.[26] These models, however nascent, invite studies of human sociality into scientific mappings of ecosystems in a way that demonstrates the extent to which modern science is itself transforming.[27]

In its most encompassing sense, urban ecology spans a vast disciplinary terrain that requires combinations of natural and social science analytics to capture. At the same time, these combinations can be extremely dynamic. Thus in concept and practice, designing a single, universal template of the urban ecosystem is to some extent impossible; contextual specificities ensure that urban ecology will always be most appropriately conveyed in a plural sense.

To ask what urban ecology means is to recognize human social agents as natural components of circuits and habitats of exchange. It opens those habitats and circuits beyond boundaries determined only by nonhuman nature, and it forces us to grapple with the social processes through which social agents organize knowledge, power, social difference, and forms of domination.

Multiple urban ecologies thus complicate the already complex science of earth systems. We ask how and when social change takes place, and how and when material and symbolic experiences of the environment are fundamental to the form and content of change. While Auyero and Swistun (2009) explore these questions ably in their recent ethnography of environmental degradation, social action, and the experience of pollution in Buenos Aires, the challenges of integrating detailed analyses of social change and detailed analyses of biophysical change remain. This book, like the work of Auyero and Swistun, falls admittedly short of fully meeting that challenge, but it recognizes in its contribution to understanding of the social experience of urban ecologies an ultimate need to fuse this understanding with equally sophisticated treatments of biophysical transition.

As Auyero and Swistun show, and as I will also suggest, ethnographic evidence indicates a constant and largely unresolved tension between the production and control of "valid" knowledge—that is, "facts" about biophysical processes—and the meanings that are attributed to the everyday life realities shaped by those same biophysical facts. We are concerned with the production of meaning because it is closely associated with human agency. It is the meaningful account of reality, after all, that motivates purposeful human action. The question of what urban ecology means, posed more clearly below, is thus an extremely important entry point for understanding how and why individuals and collectives engage the environment and one another. It is reasonable to expect that purposeful action on behalf of the environment or social justice—or both—may be driven by processes not captured or conveyed through scientific facts.

Nevertheless, scientific approaches to problem definition and solution are uniquely valuable for grappling with environmental change, and in this sense scientific knowledge usually assumes a primary, authoritative place among the many forms of knowledge through which people actually know and engage the environment in social life. Among other things, scientific knowledge enables the abstractions possible only for those ways of knowing that aspire to global, universal relevance and function, making science a key "universal" (Tsing 2004:7).

But, as Anna Tsing has shown, the abstract generalizations that universals enable are always themselves aspirations; they are "always an unfinished achievement, rather than the confirmation of a pre-formed law" (2004: 7). All universals must anchor to, and engage, local knowledge, histories, and cultural assumptions, and from that engagement they identify the knowledge that can move across and between unique social contexts. As Tsing notes, "Whether it is seen as underlying or transcending cultural difference, the mission of the universal is to form bridges, roads, and channels of circulation" (ibid.).

The challenge, perhaps, lies less in the identification and circulation of the universal—in this case, scientific knowledge and urban practices rendered as ecology—and more in the problem of its limits. Often, "those who claim to be in touch with the universal," Tsing writes, "are notoriously bad at seeing the limits and exclusions of their knowledge" (ibid., 8).

To discern those limits, we must position our inquiry in a way that looks for the particular ways and moments in which universals are used, for in practice, as we shall see, they are never politically neutral. Tsing proposes the analytic of the "engaged universal" to capture the dynamics of universals in practice, and it is from here that we may consider urban ecology as a bundle of processes that elaborate and anchor the science and practice of urban ecosystem ecology and management to a social and political context specific to Nepal's capital city on the eve of the twenty-first century.

CHASING THE MEANING OF URBAN ECOLOGY

Let us approach urban ecology, via the Bagmati and Bishnumati rivers, as an environmental problematic, by engaging the dual and simultaneous crises of environment in Kathmandu and urbanization in the global South. To do this, we engage the city, the environment, and the interactions between them as being in many ways shaped by social relations of power and meaning. Such interactions can be highly contextual, and so a singular urban ecology is never fully automatic or preconfigured.

Nevertheless, the existence of multiple urban ecologies in this sense does not preclude the fact that certain ideas about human-environment relations in the city are prioritized over others in powerful fields of action like policy or environmental management. We notice these and try to mark moments when specific conceptualizations of urban ecology converge, challenge one another, and change. This approach foregrounds the ways that cities and their socionatural environments are mutually produced over time. How are presumed intersections between environments and the urban context made, and made *meaningful*, in specific settings? Here, making the environment has simultaneous material articulations (through natural resources, for example), and more overtly social ones, like historical processes that influence ideas and imaginaries of nature itself (Williams 1980; Grove 1989; Peet and Watts 1996; Raffles 2002).[28]

I focus largely on the social and political dimensions of urban ecology in Kathmandu, at a time of profound political change. Such a focus does not imply that Kathmandu's environment, or any other, is a wholly social construct. Social constructionist approaches to the nonhuman world tend to emphasize how nature is socially produced through discourse and practice. While this may enrich our understanding of the ways that social processes structure perceptions of nature and knowledge, it can also fail to "recognize that not everything is (or can be) socially produced to the same degree, and it may overestimate the transformative powers of human practice" (Peet and Watts 1996:267). At the same time, as I will recount in this book, biophysical approaches that fail to take social and cultural context into account fall equally short.[29] Both are needed, and an analytical balance, though perhaps never perfectly achieved, must always be sought.[30] In the case of the Bagmati and Bishnumati riverscape, as stated previously, a wealth of biophysical data and policy analyses clearly existed, but the persistence of degradation suggested a need for more attention to the complex social interactions through which the problem was framed and engaged. In this sense, urban ecology in practice was itself a key component of the overall problem of river degradation in Kathmandu.

In the summer of 2000, in Kathmandu, I requested an informational meeting with the Nepali director of a prominent NGO. This director, a successful development professional from an ethnic group underrepresented among the city's NGO leaders, was considered to be an authority on urban development and change. We met in his large office, packed with towering piles of documents and policy reports, in a cement bungalow in Patan.

Our meeting began with a verbal tracing of mutual contacts active in Kathmandu's development and conservation organizations, mapping our positions in the social and professional landscape of "development"—both a process and an industry—in Nepal. The director asked me to detail the research project that I was designing at that time. I described my interests in urban development in Nepal, which dated to my first visits to the city in the early 1990s. That had been a period of change and anticipation in the capital; it fell on the heels of the People's Movement (*janā āndolan*), which had brought about the reinstatement of democracy and Nepal's first free elections since 1959. During that period I developed interests in the demographic, environmental, and political changes that followed in democracy's wake. I explained to the director that I intended to study conservation and development initiatives along the urban reaches of the Bagmati and Bishnumati rivers to try to make sense of river degradation and the impacts and politics of a series of failed river improvement initiatives.

Upon my first use of the phrase "urban ecology," the director interrupted me to convey, with excitement, his delight in the work I was describing. His organization, which since the early 1990s had designed and implemented civic education programs in the Kathmandu Valley, had recently decided to shift its programmatic focus to urban ecology and urban environmental change. But just as he finished, the director paused; his expression changed. Finally he asked, "Can you please tell me what urban ecology *means*?"

I was taken aback at first by what seemed to be the irony of the question. What urban ecology meant in its Kathmandu context was precisely *my* intended analytical puzzle, the fundamental question embedded in the research agenda I had just outlined. I had approached the director as an authority in the matter, but in the space of a question, he reversed the inquiry, leading us both to address how our own actions, our own professional activities—indeed, our own politics—might also have something to do with what urban ecology meant. I was immediately reminded of the authoritative knowledge that my position as a foreign researcher with contacts in global and regional development organizations almost automatically conveyed, regardless of my intentions or the level of knowledge I considered myself to possess.

But the director's question also made clear that urban ecology was more than a mutual research interest or a category that captured our shared interests in urban settings and environmental conditions. It was also a form of authoritative, "expert" knowledge laden with powerful potential. As the conversation continued, we were mutually forming, assembling, and establishing the kinds of knowledge that would constitute urban ecology as a legitimate basis for the specific forms of action in which we took interest. This process was making urban ecology into something about which we shared an understanding, and around which we could conceivably organize mutually intelligible practices.[31]

But the director's question had no immediate, singular answer. Although he and his colleagues had access to development publications that outlined what urban ecology "meant" to particular institutions and potential donors, and although they attended conferences, workshops, and trainings during which the concept was unveiled, explained, defined, and diagrammed, my presence, perhaps, offered an unusual ethnographic opportunity both to discern urban ecology as a potentially lucrative trend among development donors and to engage a researcher for whom urban ecology was itself an object of analysis. It offered the possibility of contestation, of questions and answers, and of exchange in the form of dialogue, however asymmetrical the dialogue might be.

Likewise, while I could read biophysical accounts of the state of Kathmandu's rivers or review the specific development projects and policies intended to improve the riverscape, the answer signaled by the director's question—Can you tell me what urban ecology *means?*— could not come exclusively from such documents; it was an answer, in fact, that most finished development reports regularly erased or backgrounded.

The question between us, and the way the question was posed and counterposed, underlined urban ecology's constant production and contestation and also its considerable power and urgency. In this case, which I would leave our conversation and go on to study, Nepal's historical moment, geographic and sociopolitical context, and even the positions of individual actors would all have bearing on the "meaning" of urban ecology and the practices through which it was enacted.

My interest in articulations of what urban ecology *meant* to a range of informants, and their assertions of the *implications* of those meanings, led back to the riverscape in ways that foregrounded discourses circulating on a stage set by activists, NGO workers, and government bureaucrats. These groups drew the ethnographic field outward from the Bagmati and Bishnumati, and into metropolitan, regional, and global circuits and networks—and then, often, these lines of inquiry reversed direction, turning those circuits and networks, in Annelise Riles's terms, inside out.[32] While the urban reaches of the Bagmati and Bishnumati constituted the physical field site, then, it was the exchange of information and power associated with urban ecology that unbounded that same field and drew both local and extralocal boundaries around environmental affinities and actions (Clifford 1997).

In this book my study is therefore best engaged as a multiscaled and multisited account, guided by Marcus's (1995) call to "follow the conflict." Doing this led me to trace intersecting, but territorially disparate, sites, even as my own physical presence was largely limited to Kathmandu and its riverscape (Marcus 1998).

I have organized the ethnography around constructions of, and conflicts over, the meaning and stakes of urban ecologies in Kathmandu. These hinged in part on discourses and gestures of affinity across markers that otherwise produced and reproduced social difference; such markers included caste, class, and ethnicity. I show how urban ecology practices in relation to the Bagmati and Bishnumati riverscape often involved making meaningful collective imaginaries that could be supported by logics of shared history, interpretations of the present, and ideas about the good and proper future. Throughout the book, I note specific moments when environmental integrity, as both a moral and political cause, was said to transcend other lines of social difference or to create new ones; I do this by highlighting ideas of belonging, collectivity, and moral social order. This suggests that urban ecology practices form an important arena for understanding how actors imagine social, and socionatural, coherence in the city of their future.

My approach to urban ecologies departs, somewhat, from the rich tradition of scholarship that has addressed Himalayan moral and social order through sacred landscape studies. These demonstrated clear analytical connections between historical urban forms and contemporary political power and spatial meaning. In the present work, I emphasize developmentalist logics of morality, those concepts of the desirable society that emerge and travel over intersecting global, regional, and local circuits as ideas and practices of modernity. Whelpton (2005:173) and others have argued convincingly that, in the second half of the twentieth century, developmentalism in Nepal "took on something of the status of an established religion," assuming the force of a moral discourse and promising a longed-for bridge to modernity that was so important for understanding urban life in contemporary Kathmandu. In this study, I use Whelpton's observation to analyze ideas of nature, aspirations for environmental change, and the politics of urban ecological knowledge in Kathmandu. By approaching urban ecologies through developmentalist logics of morality, I extend the robust anthropological literature on urban space, sociality, and sacred geography to Nepal's historical present, characterized as it was by political volatility, nationalist development, and globalist modernization.

Although it is a departure from the analytical frames of past studies, my analysis of urban ecologies builds outward from the formative work of scholars such as Neils Gutschow, Mary Shepherd Slusser, and Robert Levy, who have shown how some of the historical built forms of the Kathmandu Valley trace the pattern of the mandala, the "ritual diagram with a principal deity at its center and the other divinities of this deity's retinue arranged around it" (Hutt 1995:229). Hutt writes, for instance, that the mandala was "the model for the design of Nepal's square pagoda temples and also, though less obviously, for the layout of the royal cities of the Valley during the Malla period" (ibid.). Levy suggests that this spatial pattern should be understood as a "representation of a boundary . . . within which 'ritual' power and order is held and concentrated. The circumference of the mandala separates two very different worlds, an inside order and an outside order, and suggests the possibility of various kinds of relations and transactions between them" (1990:153).

Slusser suggested that mandalic patterns across the Kathmandu Valley had important implications for social organization and hierarchy. She wrote,

> At some time there does seem to have been a conscious attempt to bring into conformance with the vastu-sastras the social and religious structure of the towns, particularly the capitals.... [The] palace occupied a large central area, as ordained in the vastu-sastras.... High castes tended to cluster around this exalted nucleus, the lower castes lived progressively further away, and, outside the wall, were the outcastes. Finally, well beyond the city wall lay the realm of the dead, the *smasana* (Nepali, *masan*), the various cremation grounds and ghats. Superimposed on such human ordering were other orderings related to the divinities. These were in the nature of mystic diagrams, mandalas in which particular sets of deities were linked in concentric rings of protection inside and outside the city. (1982:94)

The cultural traction and logic of the mandala surface throughout this book, and my preliminary attention to them suggests an importance not fully captured here. There is clearly room for more direct further inquiry in this regard.

For Gutschow and Kolver (1975), meanings and practices associated with elements of the built environment—specifically the temples and shrines in the Kathmandu Valley city of Bhaktapur—were "deeply rooted in the citizens' consciousness." They wrote, "Often, we can show that the distribution of sacred buildings is anything but fortuitous. Topographical data, social and economic facts, and the meaning that religion has assigned to them, are fused into a unity. Space, functions, and systems of their employment and interpretation form an indivisible complex" (16). Such a complex was then analytically extended to sociality, giving basis to claims such as Shepherd's, regarding the Valley's Newar ethnic group: "One lives and moves always within a series of ... mandalas: the Nepal mandala (the Kathmandu Valley), the mandalas of the cities, and the mandalas of house or temple. Conscious of being surrounded by these forms, it is not surprising that Newars tend to reproduce that form in all aspects of their lives ... the order of the mandala has also become as habitual for Newars as linear constructions have for us" (1985:103, quoted in Gray 2006:17). Similarly, Levy (1990:2) argued that forms of social order derived from, and interacted with, built space in the Kathmandu Valley. He wrote that "the elaborate 'religious' life of [Bhaktapur], the system of symbols that helps organize the integrated life of the city so that it becomes a mesocosm, create an organized and meaningful world intermediate to the microcosmic worlds of individuals and the culturally conceived macrocosm, the universe, at whose center the city lies."

Following this, Levy and others have made important claims about the relationship between spatial form and ideal social and civic orders in the Kathmandu Valley (e.g., Levy 1990:9). Their accounts nest into a broad literature on space and cosmology in South Asia, which has broadly contended that structures, cities, and regions in India and Nepal historically articulate principles found in the *Vastu Shastras* and therefore may be understood as spatial articulations of Hindu ideology.[33] As with the mandala, discourses and imaginaries of an ecologically vital city driven in part by Hindu conceptualizations of morality appear throughout this book. At times these assume centrality and at times they do not, but they are nevertheless present and ripe for further, more rigorous attention than I can provide in this work.

But while sacred landscape literatures offer important historical and interpretive detail for understanding Kathmandu as being, in certain ways, an "ethnic city,"[34] these researchers' focus on uncovering traditional ordering systems and tracing enduring complexes presents, in some ways, a challenge to those who seek to describe lived moments and processes of social change. This is particularly true in Kathmandu's experience of developmentalism and modernization.

One can read a tension between tracing spatial continuities and engaging change in the literature itself. Slusser, for instance, wrote in 1982, "If the psychological climate of the Kathmandu Valley has resisted the impact of the outside world, so too has its physical appearance. Progress, it is true, has wrought undeniable (and unforgivable) changes, but the Valley is still today a palimpsest whereon ancient designs are clear." She continues by briefly acknowledging emergent practices of modernity, describing them as a "veneer" that is embedded within a city that is "fundamentally an antique" (15, 95).

Such an approach might interpret modern, developmentalist change as automatically invasive or disruptive of traditional patterns and the social continuities one might discern through them. Development has thus, in a sense, been analytically backgrounded, or even resisted in this literature as being disordering. Slusser, for example, wrote in a time of increasing urban development that "even within Old Kathmandu, the most changed of the three cities, the increasing number of three and four story concrete houses has not yet been able to disrupt the traditional harmony of the Newar town" (1982:15). Here, a focus on tracing tradition had the potential to obscure the ways that tradition itself may be reworked and refashioned in the course of living and making modern, developmentalist urban life. Thus the sacred landscape literature provides only a partial foundation for understanding the making and remaking of modern ecology in Kathmandu. My strategy in this book, then, is to foreground the processes, be they social, ecological, or economic, that Slusser's new concrete houses signified.

Those same houses, after all, lead us directly to the capital's rivers. With concrete housing came increased demand for cement and the sand used to make it. The source of much of the sand used in construction in the Kathmandu Valley at the time when Slusser wrote was the city's riverbeds—including the beds of the Bagmati and Bishnumati rivers. The results of intensive sand extraction are eventual morphological change and a significantly altered urban riverscape. So while Slusser was quite aware of the contemporary changes taking place in the city she studied, my work develops analytics that can more fully engage development, political transformation, and ecology's contemporary strategies and practices.

Thus, as is inevitably the case, I enable certain observations and obscure others. There are fruitful and necessary ways that the questions of continuity raised in the scholarly tradition of the sacred landscape can and must engage with urban ecology in contemporary Kathmandu. Gutschow and Kolver, for example, gestured to this potential when they described their 1975 study as, in part, motivated by concern that traditional systems were endangered, and the work at that time went on to express the hope that development practitioners would learn about, and then incorporate, traditional systems into modernization projects. In a study of urban structures and transformations in Lucknow, India, Keith Hjortshoj (1979) showed convincingly how historical "forms of order in Indian cities other than those represented by modern political institutions or legal codes" influenced design and growth patterns in particular residential localities in ways neither "governed by centralized political institutions" nor "haphazard or chaotic" (8). Hjortshoj demonstrated how historical and ethnographic perspectives could be analytically combined to understand patterns of social and cultural diversity and the ways they were reproduced in concert with political transitions and the growth of modern public institutions (ibid.:11).

By combining selected elements of Hjortshoj's approach with a particular interest in the simultaneously grounded and traveling nature of urban ecology in practice, I focus on the way that actors reworked and retold Valley history, social organization, and political order as they promoted what they considered to be ecological outcomes. I address, for instance, the mandala less as a spatial artifact and more as it resurfaced, and was refashioned, in the context of modern political projects and environmental protest. I describe the relocation of old riverscape architectural features and the imagination of new ones, toward emphasizing how longstanding symbols were invoked in ways that made them modern and how environmental embodiments of forms of political and ecological organization that were represented though rarely extant became wholly new.

I aim to contribute a study of contemporary Kathmandu as a modern city-in-the-making, in which development looms large and political change is an omnipresent frame for engaging the urban ecological present and future. I employ, then, an analytical approach that emphasizes the construction of ecological affinities, urban ecology in everyday practice, and the making of moral orders and landscapes linked directly to logics of modern, developmentalist urban environments.

The idea of ecological affinities, that is, the unifying potential of shared environmental meaning, is by no means new, but it is most often noted in global arenas.[35] In turn, global assumptions about the power of ecology to transcend human sociocultural difference have been roundly and rightly critiqued for their frequent failure to see the very real cultural and political differences that characterize grounded social life. A long legacy of scholarly attention to the so-called global and local conundrum emphasizes, among other things, the ways that environment-development initiatives conceived in global arenas nearly always reproduce, or newly erect, social boundaries that prevent the very unity that global imaginaries seek to compel (Brosius, Tsing, and Zerner 1998; Forbes 1999; Peters 1996; Rademacher and Patel 2002).

My focus on modern, developmentalist urban ecologies foregrounds spheres of activism and the particular actors who took part in development and environmental policymaking through what Appadurai (2001:22) has called "the politics of engagement." As Appadurai notices, in its contemporary form it is this sphere that often affords interactions between traditionally opposed groups. Encounters between nongovernmental advocates and activists (and the mediated voices of those whose interests they claimed to represent), Nepali bureaucrats, and international development officials constitute the sphere of engagement within which official logics and practices of urban ecology were forged and reproduced in the Kathmandu case.

No study can capture all facets of environmental change in the capital, and Nepal's political transformation continued beyond the conclusion of my fieldwork for this book. This book is best read, then, as an analytical reflection on events of the contemporary past, rather than as an itemization of the country's and the city's most recent political developments. Political life in Nepal is layered, dynamic, and complex, and readers are referred to the wealth of recent literature on specific topics—including the Maoist Movement, the rise of ethnic political parties, and contemporary political history—to supplement the river- and environment-centered narratives recounted here.[36]

While this book engages dual processes of environmental and political change, the reader will also notice that conventional details about civic political practice, such as party affiliations, particular political figures, and explicitly political meetings, are missing. The intention here is to treat political transformation less as an inventory of political personalities and parties and more as a social process guided by specific state (Hansen and Stepputat 2001) and ecological imaginaries (Peet and Watts 1996). I therefore recount informants' characterizations of an environment and a polity whose structure, symbolic power, and resilience conceptually transcended the personalities and parties that animated political change in Nepal at the close of the twentieth century and early in the twenty-first.[37]

Readers familiar with contemporary scholarship in the political science and anthropology of Nepal will also notice the relative absence of attention to the rise of specifically ethnic politics or to ethnicity as a focal analytical lens. The recent proliferation of ethnic political parties in Nepal has rightly drawn significant attention, and it is appropriately engaged in the Bagmati and Bishnumati case. However, ethnicity per se, and in particular the Newar ethnicity shared by interlocutors occupying positions across the full range of this study — including advocates for the landless poor, cultural heritage restorationists, government bureaucrats, and members of my domestic sphere — are not this book's organizing analytic.[38] Instead, I recount informants' efforts to promote ideologies of belonging according to explicitly rendered ecological logics. By considering ecology in practice, I concentrate on moments when ideas of shared history, logics of belonging, and conceptualizations of the present and possible future were explicitly organized through ecology.

At the same time, Newar identity is undeniably relevant to any study of territoriality and affinity in the Kathmandu Valley, as "Newars like to think of themselves as the indigenous people of the Kathmandu Valley."[39] Since Newars are generally regarded as the original builders of the Valley's three historic royal cities — Kathmandu, Patan, and Bhaktapur — all political contests described in this book can be understood in part as collective efforts to salvage the environmental integrity of a territory that has stood in different ways over time for Newar identity even as it has also long stood as the symbolic and bureaucratic heart of the Nepali state and the Shah kingship. Although I do not offer such a study here, there is rich potential for work that more directly analyzes contemporary Newar identity formation and transformations in Kathmandu's environmental situation and built landscape.

For those particularly interested in anthropology of the environment, this book responds to calls for ethnographic accounts of the social production of urban nature in specific political contexts (e.g., Braun 2005; Gandy 2002; West 2005; Zimmerer and Bassett 2003). As elaborated above, the book is organized through a riverscape and explores the networks of those who voiced their concerns about that riverscape in spheres of activism and under conditions of extreme political volatility. While a robust body of anthropological scholarship has explored how conservation proceeds under strong states (Agrawal 2005; Greenough and Tsing 2003; Saberwal 1999; Sivaramakrishnan 1999), few studies have yet addressed how urban environmental improvement proceeds when state power is unsteady and volatile. And while scholars such as Ben Campbell (2003) have noted that, over the past several decades, developmentalist programs aimed at reversing environmental degradation have become increasingly important dimensions of statecraft in general, and in Nepal specifically, few have considered the environment–development–state making nexus in Nepal's urban context. I aim to contribute, then, to our understanding of these issues and to changing configurations of urban nature, modern life, and forms of rule in a contemporary capital city.

Finally, the reader will notice the persistent presence of the researcher—my voice—throughout. This narrative approach is meant to underscore, as many informants described in this introduction also did, the degree to which I myself was present and implicated in the very processes that I studied. Recognizing that one can never be fully cognizant of the ways that one's encounters in social life affect or alter a cultural or political situation, I have placed myself in the retelling of the ethnographic experience. My voice thus stands as a constant marker of the impossibility of purely objective study and analysis, and it is a mediator through which my interpretations are most usefully read.

1
Creating Nepal in the Kathmandu Valley

IN THE HEART OF KATHMANDU's old urban center, the braided, seasonally shifting flows of the Bagmati and the Bishnumati rivers converge. Depending on the season of the year, the rivers may swell and churn with monsoon rains, or they may barely form a trickle. Depending on your route, you may find the banks and riverbed laden with foul-smelling heaps—mounds of garbage flecked with colorful plastics, shiny chemical slicks, and discarded consumer goods that persist beyond their utility or fashion. You may find the banks host to human settlements: some are the makeshift huts of those just arrived or just barely surviving; some are the more permanent homes of those who have saved enough to build shelters of brick or concrete on the eroded riverbed; still others are the imposing mansions of the urban elite. Settlements give way to temples—some crumble under the weight of passing time and neglect, while others are alive with the sounds, colors, and meaning of ritual activity.

The wastes of contemporary urban life, the settlements of thousands of urban residents, and the vast ancient templescape that line much of the rivers' urban reaches do not fit into neatly defined territories. They overlap, stretching into and out of one another in uneven, and often unpredictable, ways. To walk the banks of the Bagmati and Bishnumati rivers is to encounter simultaneously the natural, cultural, and political history of a city and the nation-state it dominates. It is to observe grand temples, the surviving built forms of the past, and to see the sometimes desperate shanties that make up the built forms of the present. Amid both lay deposits of waste that a contemporary city has expelled in hopes that they will simply disappear. It is to witness before you the tensions between land and water, past and future, and waste and wealth being lived in this growing, changing city.

Teku Dovan, a large temple complex, marks the confluence of the

Bagmati and Bishnumati, and it descends into the river waters by a steep ladder of stone stairs. The complex is the mythological place of origin of Kathmandu, a cultural birthplace for the city. According to a *vāmshavalī*, an ancient text that ascribes pious beginnings to Nepal's early history, the then-king Gunakamadeva (ruled 980–998) founded Kathmandu, called Kantipur at the time, in accordance with a vision. In a dream a goddess instructed him to build a city at the junction of the two rivers. According to the vāmshavalī, it "was the sacred place where, in former times, Ne Muni had performed devotions and practiced austerities, and here was the image of Kanteswara devata. To this spot Indra and the other gods came daily, to visit Lokeswara and hear puranas recited." The king thus moved his court from Patan to Kathmandu (Wright 2000 [1877]:154).

Like walking the riverscape, tracing the social and natural threads that weave the ecology of this site demands flexibility; we are at once compelled to understand the present situation, in which the environmental state of the rivers seems primary and urgent but also to find the roots of that urgency in the history of the rivers and of the larger city that envelops them. Kathmandu is, after all, a point of convergence itself. Politically, it is the capital of a nation-state, and as such it has long served as the center of the state's bureaucratic apparatus. Economically, it is a locus of concentrated wealth and elite privilege, especially in relation to the rest of Nepal. Since the eighteenth century, Kathmandu has served as the most important center of political and economic power in the country. It has also been the seat of the monarchy. In short, the city is in many ways the center of material and symbolic power in Nepal.

Yet by the beginning of the twenty-first century the Bagmati and Bishnumati were regarded as almost intractably degraded. Environmental conditions declined as political dissatisfaction intensified, and the two rivers were increasingly regarded as being connected in some ways. Engaging environmental degradation inevitably required attention to broader debates about the political past, present, and future of Nepal. The question of the environment, then, turned on a question of the polity, and how change could happen within it. Would the monarchy, the international development apparatus, the democratic

Parliament, or some other entity eventually rise to the enormity of the problem and reverse river decline? Just who held the power to make change in Kathmandu, and when and how would that power be mobilized?

These questions point to the nation-state's history, in both the capital and the larger territory over which it held sway. By beginning this inquiry with a focus on power, specifically the power to enact environmental change, we move away, for the moment, from the biophysical features of the riverscape and toward the people who organize it and, in so doing, organize themselves. To do so is also to note historical continuities between contemporary concentrations of power in the capital and longstanding processes.

MANDALA SPACE AND POLITY

In fact, political and symbolic power have long radiated outward from the center that is the Kathmandu Valley and from Teku Dovan within it. I return to the idea of the mandala for conceptual orientation in a historical moment riddled with turbulence, dissatisfaction, and environmental decline. Recall from the introduction that various schools of analysis engaged the form and symbolic meanings of Valley architecture through the conceptual and diagrammatic dimensions of the mandala. This work extends from the German tradition of approaching the city as a unitary built system (e.g., Gutschow and Kolver 1975; Mary Slusser's famous cultural historical work, *Nepal Mandala* [1982]; and many others). Much related scholarship develops analytical associations between cartographic space and the organization of political and social life; scholars use the mandala primarily as a means for understanding how the physical spaces of the Kathmandu Valley's cities were historically arranged, in concentric circles, along a gradient of socioreligious purity and impurity (Bledsoe 2004:5). As I suggested previously, the idea of tracing power as its concentrations radiate outward lends important insights to an analysis of the politics of improving the Bagmati and Bishnumati rivers in Kathmandu, and also to understanding Kathmandu in the broader nation-state of Nepal, and, at key historical junctures, Nepal in the larger South Asian region.

Slusser, in particular, famously emphasized the relationship between historical Kathmandu and a mandala, in terms of urban design and social organization. Prior to Prithvi Narayan Shah's territorial conquest and creation of what is now called Nepal, the term "Nepal" referred only to the Kathmandu Valley itself—then called the Nepal Valley, or Nepālmandala. Slusser and others showed that, in part, the organization of Nepālmandala inscribed caste, and related prescriptions for political harmony, into geographical space.[1] Centered on the king's residence, concentric and outward radiating circles organized the ancient polity along a discernible gradient of descending ritual and political rank.

Recall from the introduction that Slusser's work and the scholarly traditions that have followed it underline the utility of the mandala as a window on the spatial and symbolic history of power relations in the Kathmandu Valley. This utility, however, and the precise historical uses and meanings of the mandala in the architectural and physical context of Nepal, are points of ongoing scholarly and popular debates.[2] These debates are often framed by the very politics suggested by invoking the mandala itself; both the place and concept of Nepālmandala are recounted sentimentally by modern Newar intellectuals as a way to recall the Malla era, when they imagine a lost period of Valley unity. Although it is unlikely that perfect political and social harmony existed at that time, present-day nostalgia for the Malla era plays an important role in contemporary identity formation for the ethnic group that claims indigenous status in Kathmandu Valley, the Newar (Bledsoe 2004:60).

In my study, the mandala reappeared in the practice of urban ecology through state spatial practices and performances of citizenship. As I will describe in more detail later in the book (chapter 5), the emergency and *loktantra* periods witnessed specific uses of, and encounters with, a physical mandala constructed in a new urban park. On this physical mandala, however, groups from Nepal's furthest sociopolitical margins gathered to articulate and amplify political demands. Citizens who occupied it were from Nepal's furthest margins and were contemporary "outcasts" in social, economic, political, and a host of other ways. Yet in the turmoil of political transformation, they took over—both physically and profoundly symbolically—the very center of Kathmandu's Maitighar mandala. In doing so, they inverted the historical power relations suggested by the mandala, remapping (through the practice of urban ecology) old landscapes of caste position and political power. They conveyed, therefore, the tremendous extent to which the social order implied by the mandala was itself transforming. Thus, the processes that we must grapple with to fully address environmental change in Kathmandu involve even the very social and political changes that eventually led to the occupation of one of the state's own modern mandalas.

Before proceeding to that case, however, and before addressing who, in the end, would reign over the rivers (and in so doing assume symbolic and active power to make positive change) we must review the legacy of those who controlled the polity and its landscape in the past, and we must consider the enduring effects of that control on the politics of contemporary river restoration in Kathmandu.

BUILDING MONARCHY, DEMOCRACY, AND NATION

To understand legacies of political power, I begin by briefly historicizing the political organization of the Kathmandu Valley over time, addressing the key elements of monarchy, nation building, the rise of an international development apparatus for Nepal, and democracy. I do so with studied caution, however, since every retelling of history is fragmentary and is animated primarily by the voices of those powerful enough to have made themselves heard as history makers. My discussion here is in no way immune to the partiality of historical narrative; the form and content of national history are actively contested among scholars and Nepali citizens alike, perhaps never more fervently than in the contemporary present. A great deal of Nepal's history was silenced in the more conventional, and historically dominant, accounts to which my tracing of history will refer. There is much to recover. Limited though they are for gleaning a full and detailed history of Nepal, conventional accounts do provide useful anchors for making sense of Kathmandu's political and environmental present at the turn of the twenty-first century.

To begin, the first dated reference to Nepālmandala is found in seventh-century inscriptions from the Lichhavi Dynasty;[3] these references are followed by mention of a sacred realm, or *deśa*. References to Nepāla Deśa, or Nepālmandala, signify the Valley as a Hindu realm, which Burghart (1984:104) describes as "an auspicious icon of the universe centered on the temple of the king's tutelary deity [Taleju] and demarcated on the perimeter by temples—often four or eight, which were situated at the four cardinal directions or eight points on the compass."

A templescape formed the boundaries of the deśa, designating the space inside as being pure and the space outside as being impure. Burghart refers to Hamilton's (1819:192) description of a Brahmanical scheme with fifty-six universal deśas in the Sacred Land of Hindus. Nepālmandala's sacred landscape contained 5,600,000 *bhairav*s and *bhairavī*s—male and female spirits of Shiva and Shakti. Kings in the Malla (1200–1769) and Gorkhali-Shah (1769–1990) eras acted symbolically as protectors of this deśa.[4]

Wealth from trade provided the material basis for three major city-states in the Valley: Kathmandu, Patan, and Bhaktapur.[5] At first a single kingdom, the three became autonomous following a period of complicated Malla succession between 1484 and 1619. While they prospered, the later years of the Malla era saw these city-states increasingly marred by rivalries that weakened the Valley and made it vulnerable to conquest by the "Great Unifier" from Gorkah, Prithvi Narayan Shah, in 1744 (e.g., Rai 2002).[6] Kathmandu, Patan, and Bhaktapur did not unite against Shah, whose campaign in the three cities of the Valley was completed by 1769.

The conquest of Kathmandu was a central part of a general Gorkhali campaign of regional expansion. Prithvi Narayan Shah based his new court in Kathmandu, claiming kingship over all three Malla kingdoms. The throne of the Shah Dynasty was brought from Gorkha, and the new king assumed his place in the compound of the deposed king (Burghart 1984: 111). This moment fixed the Nepal Valley as the core of power and authority for the Gorkhali nation-state; at the same time, it relegated the territory outside of the Valley to the margins of political power.[7]

In the early period of the Shah kingship, former designations of who could live where in Nepālmandala were blurred. In particular, the lowest caste groups that were previously forbidden to live inside the city were joined by social groups once found exclusively inside it.[8] A

Newar ritual that traced the city walls during the Malla era (*upako va-negu*—Newar for "walking the town") was adapted by Gorkha rulers as *deśa ghumne*, performed annually during the Indra Jatra festival, reproducing ritually that which had been lost in physical space.

While growth and change reshaped the previous boundaries of Nepālmandala, policies related to ideas of realm purity intensified the distinction between that which was considered inside and outside it. Mark Liechty (1997) noted that, by the early 1700s, Gorkhalis self-consciously thought of their region as being distinct from, and ritually superior to, much of the territory to the south, mainly because it had remained "uncontaminated" by Muslim and British rule in much of India. For Prithvi Narayan Shah, assuming the role of protector of the realm involved maintaining the region's status as the "pure, true Hindustan" (*asal Hindustan*).[9] As territory to the south was increasingly consolidated under British colonial power, Shah rulers intensified their distinction between land inside the Valley (pure) and outside it (impure). Impurity was to be kept outside the realm, and that which was non-Hindu, foreign, or considered immoral was to be vigilantly repelled.[10] After 1817, Gorkhali rulers saw theirs as the only remaining "pure" realm in the entire region.[11]

While Prithvi Narayan Shah was not the first ruler to espouse ideas of realm purity, he enacted them through policies that would later be interpreted as the beginning of Rana isolationism. By the Rana period, the exclusion of all dealings with foreigners from public life became policy.[12] When Rana rule (1846–1951) commenced under Jung Bahadur, state and nation building started to mirror some of the European colonial policies seen elsewhere on the subcontinent.[13] For example, with the declaration in 1854 of the Civil Code, or *Muluki Ain*, a national caste system was codified and diverse ethnic groups organized according to a hierarchy of state-defined purity.[14] The Muluki Ain also replaced the Hindu Laws of Manu for forming legal judgments (Hofer 1979; Levine 1987). All peoples were assigned a *jāt* (caste) and categorized into one of five hierarchically arranged groups.[15] Whereas previously the territories that diverse peoples occupied determined their distinctiveness, the new code classed everyone according to their "species" (Burghart's translation of jāt) and assumed that they inhabited a singular and common territorial unit (Burghart 1984:117). Caste thus became a unifying tool of state making, and scheduled difference among castes became an important logic for imagining national unity. Not incidentally, this social arrangement also reproduced and sustained high-caste Hindu dominance in state affairs.[16]

Outside of the Muluki Ain, however, Rana rulers did not embark on a comprehensive project of creating a singular national identity shared by all people throughout the country. They did, however, make efforts to present Nepal to outsiders as a state with a coherent culture, and with aligned realm and territorial boundaries.[17] For example, by the 1930s, the government began to refer to its kingdom as the "realm of Nepal" rather than the previous "Entire Possessions of the Gorkha King," officially conflating the realm of the Valley with the rest of the political territory—and collectively calling it "Nepal." Later, the official language (*Khas kurā* or *Gorkhālī*) was nominally changed to Nepali, disassociating the language from the group who ruled the country (those coming from Gorkha, or the Khas people). This was less an attempt to build popular support for the language, because the Ranas didn't promote the use of Nepali nationwide, than it was an effort to present Nepal as a culturally unified nation to the rest of the world. Of particular importance in this regard were India's colonial British rulers.

Deliberate state cultivation of a Nepali national consciousness began after the fall of the Rana regime and the reinstatement of King Tribhuvan in 1951 (Burghart 1984).[18] Although certain private groups had been working for decades to foster Nepali national consciousness underground and from exile in India, it was not until 1951 that the state began an official project of nation building that involved state-run schools, the definition and promotion of specific national dress, an anthem, a national language, the formation of an official media, and other features (Pfaff-Czarnecka 1997).[19] Radio Nepal began a project that would eventually reach remote hill villages with *samāchār* (news) and musical programs intended to cultivate and convey a cultural sense of Nepaliness, or *Nepālīpan*. The use of such nation-building techniques intensified under King Mahendra, whose Panchayat government launched a nation-building campaign in its most classic Western sense. At that time, official claims to a singular, unified, unique Nepali identity were used to officially challenge and counter the legitimacy of multiple political parties, convey the idea of Nepal as being the world's only Hindu kingdom, and reinforce allegiance to a politically autonomous nation-state.

During the Panchayat period (1960–1990), the ruling elite regarded cultural diversity as a key obstacle to achieving national unity. This was clearly demonstrated with the legal "eradication" of caste and a declaration that the entire Nepali citizenry was equal under the law.[20] Caste and ethnic categories were not explicitly used in census aggregations at this time; although jāt information was recorded, census takers did not tabulate it. Instead, they counted citizens on the basis of language and religion. This had the effect of downplaying difference and emphasizing national uniformity by "demonstrating" national homogeneity through an appearance that a majority of the population spoke Nepali and practiced Hinduism (Hangen 2000:59).

Monarchical power retained some form of symbolic unifying significance throughout the twentieth century, even in periods of indirect monarchical rule. Indeed, the enduring centrality of the kingship to the conceptual coherence of Nepali national identity was underscored as recently as 1997 when Prayag Raj Sharma wrote, "Remove monarchy and there is no state, and minus the state, there is no nationalism. The only form of nationalism the ethnic groups have known about or been familiar with is in the framework of the Nepali state" (Sharma in Gellner, Pfaff-Czarnecka, and Whelpton 1997:482). At the center of that state was an enduring kingship. Yet experiments with, and advocacy for, forms of democratic governance that would share power with the monarchy punctuate the second half of the twentieth century, with roots extending long before that as well (Fisher 2000).

PEOPLE, THE KINGSHIP, AND EXPERIMENTS
WITH DEMOCRACY

In 1950, disaffection with the Rana regime and a longstanding movement for sweeping political change reached a critical juncture. In November, King Tribhuvan, who until then had been kept a virtually powerless figurehead by the Rana prime minister, fled Kathmandu and sought political asylum in India. This move catalyzed the eventual demise of the Rana regime. On February 15, 1951, King Tribhuvan returned to claim his position as the constitutional monarch of a *demo-*

cratic polity. His announced intention to hold a constituent assembly and draft a new democratic constitution secured this claim.

But political instability delayed promised constituent assembly elections, and just a few years later, King Tribhuvan died. When his son Mahendra assumed the throne, he immediately declared direct rule over Nepal's new "democratic" system. Naming his own prime minister, the new king banned all printed criticism of the monarch and presented a new constitution—drafted without the input of a constituent assembly—that allowed basic liberties (including religious freedom, free speech, and free political affiliation) and declared equality among citizens. The new constitution also reinforced monarchical sovereignty by encoding royal immunity from the law and granting the king the capacity to declare an emergency if the government should fail. The king thus remained the "protector" of the people, holding ultimate power over the polity and the military.

By 1959, after nearly a decade of delayed democracy, free elections took place nationwide for the first time. The returns brought an overwhelming victory to the Congress Party, but, citing an outbreak of violence in Gorkha, King Mahendra invoked his emergency powers and dismissed the elected prime minister in a royal coup. The king ordered elected members of Parliament and the cabinet arrested, and he declared all political parties unfit to rule. He then announced that the democratic experiment had failed, replaced it with the Panchayat system, and famously claimed that Panchayat democracy was most appropriate for the needs of Nepal. A new Panchayat constitution was introduced in December 1962; it reascribed power to the kingship and banned all political parties and associated activities. Once again, it granted the king key political powers that solidified the place of the monarch and the capital as centers of state, nation, and realm.

It would be a mistake to understand the Panchayat period as one that was free from political contestation, however. Fujikura (2001) has shown how development initiatives under the Panchayat formed conduits through which "democratic culture" developed, was disseminated, and was exchanged. While nominally partyless, then, the Panchayat era witnessed nascent new forms of political mobilization that maintained political party activity and laid a foundation for later, more overt political reform.

King Mahendra's Panchayat regime focused on two interrelated projects: One, described above, was the formation and diffusion of a unifying national identity. The second was broad socioeconomic development. "Democracy," as such, left the official stage, replaced by national consolidation and five-year development planning.

Fervent efforts to modernize and develop Nepal were organized through five-year planning schemes, which necessitated a new bureaucratic apparatus. A civil service proliferated, centrally administered in Kathmandu. Regional and international governments initiated and financed a range of development projects, establishing associated networks to channel financial resources and exercise influence on political and policy-related matters. For international governments, Nepal's geographic position between India and China made the country strategically important in the cold war's global competition between Communism and "containment." Nepal received development assistance from countries on all sides, unleashing a staggering amount of development money into the economy.[21]

Foreign aid as a percentage of development expenditure rose steadily from the First Five-Year Plan (1956–61) through the Fourth Plan (1970–75) (Acharya 1992:9). It then rose sharply from the mid-1970s through 1990. Acharya shows that national dependence on foreign assistance in 1990 was increasing, noting as well an alarming rise in the national foreign debt. By 1990–91 the outstanding external debt was NRS. 46 billion, or 46 percent of estimated GDP (Acharya 1992:10). By 1997, well over half of the government's budget came from foreign aid (Dixit 1997:175), following, rather than reversing, trends of the 1970s and 1980s in which Nepal maintained one of the highest per capita levels of foreign aid in the world (Seddon 1987).

Massive development aid financed a government bureaucracy designed to manage development projects, and the result was exponential growth of the urban-based state apparatus (Fujikura 1996). Dixit and Ramachandran (2002) and Blaikie, Cameron, and Seddon (1980) and others have shown how this growing bureaucracy absorbed the Kathmandu Valley's most educated citizens. In so doing, they argue, development played a part in muting potential political opposition. Sharp increases in foreign aid from the mid-1970s onward corresponded with a shift from bilateral to multilateral aid agreements, and, "with massive amounts of multilateral foreign aid flowing in, one of Kathmandu's growth industries [. . . became] the development of layer upon layer of international and local nongovernmental organizations (INGOs and NGOs). Employing hundreds of expatriate and thousands of Nepali middle-class professionals, these NGOs systematically drained off aid dollars in a complex system of legitimate needs ... and illegal practices" (Liechty 2003:49).

As development aid poured into the country, it created, in part, its own self-perpetuating structure and requirements. Although funds were often designated for rural development purposes, "by the time the money filtered through the maze of centralized bureaucratic bodies and their affiliated nongovernmental organizations, often very little [remained] for projects at the 'grassroots'" (Liechty 2003:49). In short, development had generated systems, interests, and a professional class that stood apart from the poverty that it was designed to alleviate.

Nanda Shrestha (1990:18) estimated that only 10 percent of aid during the Panchayat era actually reached the rural poor, while 30–40 percent went directly to the royal palace. In fact, after democracy was reinstated in 1990, many new NGOs voiced explicit critiques of Panchayat policies that had required all development funds to be channeled through royal bureaucratic agencies. Notable among these was the Social Services National Coordination Council, overseen directly by the queen.[22] Liechty (2003:48) suggests that the percentage of aid that directly benefited royalty and the elite was even higher in the post-Panchayat period, citing an article in the *Independent* (from November 27, 1996) that estimated that, by the mid-1990s, a full 70 percent of Nepal's total annual budget had gone to running the national government headquartered in Kathmandu. Precise figures are difficult to glean, but it is clear that development institutions and projects figured prominently in the reproduction and eventual amplification of critiques of royal legitimacy, power, and wealth.

Despite deep public frustrations with the legacy of foreign aid in Nepal's national development (A. Tiwari 1992; Udaya-Himalaya Network 1992), Kathmandu witnessed an explosion of NGOs in the decade after the reinstatement of democracy in 1990. Often portrayed as the key to reorienting Nepal's development industry toward better coverage, results, and accountability, these organizations nevertheless remained fully dependent on continued inputs of foreign aid.

Stacy Pigg (1992) has shown that development aid also catalyzed specific social processes with important effects, including shifting cash flows and the reorganization of rural and urban relations. As the development apparatus flourished, the state became, in large part, a cash source rather than a cash sink (Caplan 1975). This facilitated new forms of movement and exchange between the capital and the rest of Nepal, and it generated new forms of competition for resources throughout the development economy. It also opened opportunities for upward mobility to more than the historical elite.

Throughout the 1990s, the development industry served as a major source of employment for the urban middle class, as well as a major source of revenue for the national government. It was implicated in the reproduction of important distinctions between the city and the countryside. This attracted the notice of some public intellectuals, who noted links between the centrality of Kathmandu and the powerful momentum of the rural-based revolutionary opposition that eventually engaged in armed conflict with the ruling government.

Distinctions between rural and urban were also reproduced in patterns of democratic change. In the winter of 1989, an almost exclusively urban movement to reinstate democracy began in the capital. The *janā āndolan*, or People's Movement (hereafter, jana andolan), saw political parties and young activists take to the streets to demand Panchayat reform. These demonstrations marked Nepal's most popular political movement to that date, and by April, King Birendra—whose kingship was, for the first time, openly ridiculed as part of the demonstrations—bowed to popular pressure and agreed to "democracy," or *prajatantra*, and a new democratic constitution. The new constitution, however, reinforced Nepal's status as a Hindu kingdom, and it reserved far-reaching powers for the king. Article 127 of the new constitution maintained the king's ability to suspend democratic processes, effectively granting him the absolute power that Birendra's father had invoked in the 1960 royal coup.

Initial high expectations for rapid, sweeping change under 1990s prajatantra gave way to extreme disappointment and disillusionment, as government policies proved ineffectual, parliamentary formations changed almost annually, and a Maoist revolutionary resistance movement gained popular strength in the countryside. Democratic aspirations that seemed to hold infinite promise in 1990 eventually became synonymous with disorder, greed, corruption, loss, and ecological degradation in the capital city (Rademacher 2007). Middle-class disillusionment with democracy, and fear of the Maoist resistance, created a sense among Kathmandu's elite and middle classes that a resurgence of royal power was necessary. Some urban elites believed that concentrating monarchical power could facilitate much-needed stability and quell a growing Maoist resistance. This resurgence assumed the form of a state of emergency, declared in November 2001. Although initially supported by many of my professional informants, the king was nevertheless expected to reinstate democratic elections once these critical tasks of order and governance were accomplished.

But months, and then years, passed. After King Birendra was murdered, King Gyanendra's absolute rule proved anything but temporary, and although he continued to consolidate royal power, he managed little discernible progress in calming the armed conflict. Support for the king among urban elites eventually dissolved. By the time he announced a complete royal seizure of power on February 1, 2005, a nascent alliance between seven major political parties and the leadership of the Maoist resistance had solidified. The force uniting these very disparate groups was their declared opposition not only to King Gyanendra but also to the Shah kingship itself.

Sustained demonstrations began in April 2006, and these were the formidable product of a united front of political party leadership and Maoist cadres. Unlike the 1990 jana andolan, this movement, which called for the reinstatement of Parliament and featured unprecedented opposition to the monarchy, reached far beyond the capital. Protests occurred all over Nepal, but the most significant and sustained demonstrations took place on the outskirts of Kathmandu, where protestors refused to relent despite curfews and violent police attacks. After weeks of unrest, King Gyanendra finally consented to reinstate Parliament; soon afterward, that same body voted to strip the king of all political powers. With this vote, King Gyanendra, the descendant of the 238-year-old Shah Dynasty, was effectively reduced to the role of figurehead. Following elections in April 2008, the future of the monarchy then rested in the hands of an elected, overwhelmingly Maoist, Parliament.

Longstanding struggles for new forms of governance, and recurrent debates about who should "develop" Nepal and how, form modern undercurrents that are important for understanding what urban ecology meant to differently positioned actors' assessments of who had, or should have, the power to make environmental change in the capital. These understandings link to rich and complex historical foundations for the form, content, and social traction of urban river degradation by the end of the twentieth century. While this chapter has managed only to sketch the outlines of an extremely rich and varied set of historical experiences of Nepal, it nevertheless points to the importance of governance and development for establishing what urban ecology and urban environmental improvement might mean in the period of transition from the twentieth century to the twenty-first.

2
Knowing the Problem

BY THE LATE 1990S, river conditions in Nepal's capital city had reached a new breaking point, amplified in prevalent news media accounts. On February 7, 2000, a *Kathmandu Post* headline declared "Billions Spent but the Rivers Still Polluted," and the article indicated that government and development organizations had disbursed over NRS 3 billion in allotments toward improvements for the Bagmati and Bishnumati rivers to almost no discernible positive effect: "Despite the expenses, Kathmandu Valley's rivers—Bagmati, Bishnumati, Dhobikhola, Hanumante and Tukucha—continue to serve as massive sewerage systems for the Valley's overall population of 1.5 million. Industrial waste is also dumped into them recklessly.... There is a total lack of accountability and coordination among more than a dozen authorities that are supposed to be concerned about the plight of the rivers." Broad public consensus held that something was terribly wrong with river water quality, flow quantity, and morphology. Actual river improvement was clearly overdue. But precisely what to do—and the exact causes and consequences of river decline—remained the subject of intense debate. River conditions were at once intolerable and yet were so controversial and complex that the problem of restoration seemed intractable.

What *was* the problem? How did the measurable, biophysical details of urban river decline intersect with the somewhat less tangible, and yet extremely powerful, cultural and political dimensions? To explore this question, I consider here various combinations of scientific and alternative ways of gauging river degradation and how their interplay produced and prioritized the very categories through which degradation was defined.

Competing definitions of degradation activated different logics of who did and who did not belong on the riverscape once it was eventually restored. In this sense, ideas and practices of urban ecology legitimized particular forms of socioenvironmental inclusion and exclusion, forms reinforced by logics of precisely who should be held accountable for river improvement, and how. Embedded socioenvironmental inclusion and exclusion, then, were competing ideas about the forms of governance and authority that would best ensure the rivers' restoration, protection, and long-term well-being. Knowing and defining the problem of river degradation was critically interlinked with questions of how social change should take place in Kathmandu.

In this chapter I review multiple, simultaneous framings of river decline in order to show how those framings competed for legitimacy and active power. The presence of multiple framings reminds us that, in concept and practice, urban ecologies are often contested in ways that prevent their full coherence across social groups. But it also compels us to attend to the stakes of the competition over what urban ecology means to differently positioned actors. In this case, the stakes were no less than entitlement to direct the social and environmental future of Nepal's capital city—its form, its "natural" spaces, and the distribution of environmental and political power among its inhabitants.

Adherents of different urban ecology frames selected certain social and environmental facts to advance as matters of primary importance. In crafting certain definitions of urban ecology, what did each frame then obscure in turn? How did the biases of each frame prohibit a holistic view of river management and river restoration, and, in some cases, render it impossible for communication or compromise across differently positioned actors who were nevertheless passionately committed to the same cause?

In their recent book *Flammable*, Auyero and Swistun (2009) consider multiple narratives and observe, as I will, elaborate plural perspectives on environmental conditions and their relationships to social well-being. They observe the multiple and contradictory positions of every relevant actor in their story, showing how profoundly positions will shift according to time, informants' aspirations, and informants' relative social positions. Powerful actors may be at once benevolent and exploitative; less powerful agents may be advocates for their well-being as they perceive it but also willfully complicit in the conditions that produce their own environmental or social marginality. Auyero's and Swistun's account shows that a collectively desired "way out" of difficult environmental situations is often uncertain, always shifting, and always better understood when analyzed in terms of symbolic violence and misrecognition. These analytics can provide tools to discern why people endure the complex forms of illness and suffering that environmental degradation often produces.

In this chapter and throughout this book, river narratives change over time and across space partly in response to a sociopolitically chaotic moment. For many river-focused actors, the relative chaos of the present shaped a vision of a future that they imagined to be coherent and ordered. I outline three frames for knowing river decline as a problem and for thinking through its possible solutions. The first is a relatively dominant, official framing based on government studies, policy initiatives, and development plans. A second foregrounds the implications of river management plans for thousands of landless poor settled along the degraded riverbanks, and a third emphasizes particular historical and cultural dimensions of degradation and restoration. These three frames were not necessarily exclusive of one another, nor were they static and unchanging as time progressed and political circumstances shifted. Rather, different ways of characterizing problems and solutions on the rivers — an official state-development narrative, a housing-focused narrative, and a culturally focused narrative — moved, along with their advocates, into and out of agreement with one another, sometimes producing collaborations and sometimes producing staunch oppositions.[1] Throughout, however, the potential power to shape the future of a capital city and a center of Nepali national identity was omnipresent. To begin, let us return to the late 1990s, when river degradation assumed a particularly prominent place in popular and official consciousness.

In the fall of 1997, I spent weeks bicycling back and forth between my host family's home in Satdobato and a large, shaded bungalow across Patan in Dhobighat. A long driveway set this white house in Dhobighat away from a street noisy with the beeping horns of passing cars, the shouts of vendors, the bustle of shoppers, and the laughter of children moving in uniformed clusters between home and school. The quiet building housed the office of an international conservation agency and, more important, its extensive library of books and reports on environmental issues.

My effort to understand urban river change in Kathmandu began in this library, seated, somewhat like the government bureaucrats whom I would later spend hours interviewing, before towering stacks of reports. Some were official studies, commissioned by ministries and missions to investigate urban river conditions or plans for future management. Others—usually glossier, with more photos and published as books rather than being bundles of paper—were studies commissioned by international development agencies, or undertaken by international consulting firms, in cooperation with His Majesty's Government (HMG). All of these texts concluded with a clear outline of problems; all offered policy prescriptions to solve those problems.

For weeks, I scoured these reports, piecing together a story of river change and biophysical assessments through which to detail it. While certain kinds of data were undoubtedly missing from the larger story these documents told, for the most part they were relatively comprehensive from a biophysical point of view. I found myself puzzling over the sheer volume of information, which emphasized the absence of any discernible progress toward reversing river degradation. The reports outlined clear interventions that would mitigate river decline; so, with the existence of so much policy-ready knowledge, how could river conditions be deteriorating so quickly and completely? How could river degradation persist when the government had spent billions, as the *Post* reported, to produce all these reports and put their plans into action?

When I posed these questions to the Nepali director of the international conservation organization that ran the library, he replied with a knowing smile. "Kathmandu is full of good reports," he said dryly, "and in fact that is most of what we do. We conduct excellent studies and write top-notch reports. Knowing the problem is *not* the problem here."

Indeed, river degradation and a comprehensive plan for river restoration were meticulously detailed in publications such as *The Bagmati Basin Management Strategy* (BBMS), the main planning document for the Bagmati Basin Water Management Strategy and Investment Program. This was a policy proposal to the national government, held in active consideration even as I pored over its attendant documents in the library in Dhobighat. Completed under the auspices of organizations including the Japanese Grant Fund, the World Bank, the International Union for the Conservation of Nature, the consulting group Stanley International, and the Nepal Ministry of Health and Physical Planning, this report became somewhat of a benchmark in official articulations of river degradation as an official policy and development problem in Kathmandu.[2] The conclusion to the BBMS listed restoration goals and policy prescriptions to accomplish everything from water quality improvement and sediment replenishment to temple complex restoration and reclamation of open public spaces.[3] While the study clearly defined biophysical degradation and restoration, the possible future for the riverscape would be conditioned by political, cultural, and environmental interventions.

Actionable biophysical data were thus not missing in the case of Bagmati and Bishnumati decline. Indeed, the problem was "known," according to the criteria set forth by the development consultants who compiled the BBMS, a team that included hydrologists, ecologists, engineers, and sociologists. Yet the conservation organization director's confident but somewhat smug response to my question suggested that something was clearly amiss. An omission or a lack of facility with a particularly important way of knowing the riverscape and the city seemed to break the link between the stacks of reports and resulting effective restoration action.

Further complicating the puzzle, the director was relatively powerful across the range of actors involved with official river management. Presumably he had the ear of policymakers and ministers; as a Nepali professional working for a major international development organization, he spoke with authority to both his foreign colleagues—who controlled significant amounts of funding—and his Nepali colleagues, who held political power. With such a plethora of data and from such a position of power, why did this director represent himself and his organization's research activities as being so powerless in the struggle against river decline?

Perhaps "the problem" needed to be recast, not through the causes and features of river degradation but rather through the meaning of urban ecology itself to differently positioned actors. How might multiple meanings shed light on the relationship between environmental change and social change, and how might these be brought together to forge a coherent and actionable agenda? It became quite clear that, contrary to the director's suggestion, knowing the problem was precisely the problem.

AN OFFICIAL URBAN ECOLOGY

Let us first explore the official narrative on which the director and his colleagues in development and government relied. This narrative drew heavily on the scientific data and policy plans held in the library where my inquiry began; it combined its biophysical river profiles with selective and strategic discussions of the historical and cultural value of the Bagmati and Bishnumati riverscape. In the late 1990s, the study that enjoyed the most prominence among official river reports was the BBMS.

Like other studies in its genre, the BBMS identified many facets of degradation, but its authors emphasized some as being most pressing. Mentioned in my own characterization of river decline early in this book, these main features were severely reduced water flow and quality, significant riverbed morphological changes, and the loss of cultural and religious values historically attributed to the rivers. Degradation was assigned four major causes inside the urban area. These were, first, the discharge of nearly all of the city's sewage — completely untreated — directly into the rivers, and, second, the widespread dumping of solid waste into the rivers and on the banks. Third, sand mining in riverbeds and banks, which supplied mortar and cement materials to a booming construction industry, had produced significant morphological change and had severely channelized flow patterns. Finally, human encroachment on the banks, floodplains, and riverbeds exposed by channelization was considered a significant factor in the degradation process. The deteriorating ancient templescape along the rivers was described as an unacceptable casualty of this encroachment.

Several other studies, outlined below, assumed prevalence in what I refer to as the official narrative of river degradation that coalesced during my field research. In general, they emphasized one or all of the aspects of concern reported in the BBMS, and each offered a solution that would bring the riverscape to a more "restored" state.

In addition to forging a relatively coherent definition of urban ecology from the studies, I saw that these plans shared a link to international development institutions. In each plan, the work of a foreign donor was to have bridged the gap between the document and its execution as an actual development project. In some instances, the bridge would have been funding; in others, it involved foreign consultants, scientists, planners, and various other international experts.

This link between the official narrative of river decline and the development industry that facilitated official research, reports, and associated restoration projects attaches its overall degradation problematic to a particular idea of the appropriate future for the capital city. Its templates for river change assumed that the polity would remain organized around the scenario of state-development interdependence referenced in the previous chapter. Future distributions of political power were assumed to be largely consistent with the present; while the form of government may continue to "democratize," the ultimate relationship between the international development apparatus and the state was expected to endure and self-perpetuate. Thus, official environmental diagnoses failed to problematize governance itself, and the specific relationship between international development and the Nepali state to which proponents of political change often pointed when they read "Billions Spent, but the River Still Polluted."

Of the major river restoration and improvement projects between 1991 and 2003, the following are considered in the present study:[4]

— The Bagmati Basin Management Strategy (BBMS) for the Upper Bagmati Basin. Its document was released in 1994 and had major support from the Japanese Grant Foundation and the World Bank, but the project was not implemented. It proposed a fourteen-point action plan that called for a new sewage treatment scheme, solid waste management improvements, various effluent controls, and the "re-establishment of non-destructive historic and cultural riverside usage" (Stanley International et al. 1994:14) through the Teku-Thapatali Project (see below). Its ecological restoration plan featured a series of weir dams to raise the riverbeds and widen flow patterns of currents.

— The Kathmandu Urban Development Project (KUDP) for the Bishnumati corridor was initiated in 1991. It had major support from the Asian Development Bank, the HMG Ministry of Housing and Physical Planning, and the HMG Department of Housing and Urban Development. Its priorities

Map 2. The Upper Bagmati Basin, showing the municipalities of Kathmandu, Bhaktapur, and Lalitpur. *(Created by Katie Osbourne)*

Map 3. Major watercourses of the Upper Bagmati Basin, showing the Bagmati and Bishnumati convergence at Teku. *(Created by Katie Osbourne)*

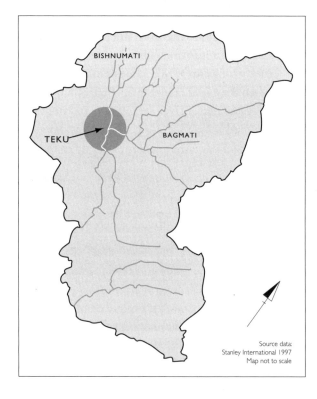

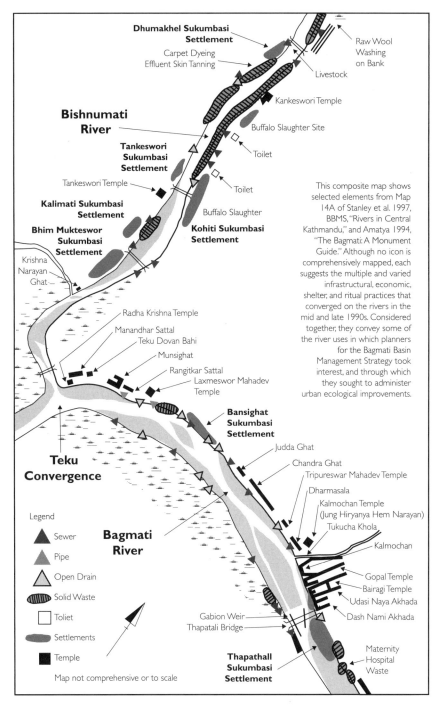

Map 4. The social life of Kathmandu's Bagmati and Bishnumati rivers.
(Created by Katie Osbourne)

were the removal of solid waste from and planting and landscaping within the corridor, public toilet construction, and a public education campaign. The larger KUDP involved road construction and infrastructural alterations throughout the city.

—The UN Park along the Bagmati and Bishnumati was planned since December 31, 1995. Its impetus and major support were from the United Nations to commemorate Nepali military service in UN peacekeeping on the occasion of the UN's fiftieth anniversary. Donor funding had not been secured during my research for this book. Its major projects were a proposed park with recreational and memorial features along a 3.6-kilometer-long strip of riverbank and the exposed bed from Shankhamul to Teku Dovan.

—The Teku-Thapatali Project at the temple complex at Teku Dovan was approved in 1996 but went unimplemented, despite support from various bilateral donors. Its major projects were restorations of temples and ghat at Teku Dovan and of public open spaces in the temple vicinity.

The official portrait of river decline included detailed biophysical data so it could be intelligible across places and audiences and could be linked to a clear and science-driven prescription for restoration. Such data are essential to any complete accounting of river ecological conditions, but they assumed a prevalence in official narratives that was sometimes de-emphasized in other urban ecology frames. The biophysical facts of degradation in the late 1990s can be summarized as follows.

Unlike the more common snow-fed rivers in Nepal, the Bagmati and Bishnumati rivers are fed by rainfall and springs. The head reaches of both rivers are located within the Valley, so surface water and waste assimilation capacity are relatively limited in the upper reaches. Water resources in the Upper Bagmati Basin are used primarily for the municipal water supply and for small-scale irrigation.[5]

River surface water and groundwater contribute to Kathmandu's municipal water system: 65–88 percent of municipal water supplies are derived from surface sources while 12–35 percent come from groundwater (the proportion from each varies seasonally).[6] The Bagmati Basin Management Strategy suggested a somewhat higher percentage of municipal supplies from groundwater, a figure of 45 percent.[7] Municipal water extraction occurs upstream of Kathmandu at Sundarijal. During the dry season, the entire river flow at this point is sometimes diverted for municipal use.[8] Drinking water quality worsens considerably on the journey from municipal extraction and treatment facilities to domestic taps (Stanley International et al. 1994:A5). Contamination of the piped water supply is common and is attributed to leaks in the delivery system. The piped water supply is also intermittent and unreliable during the dry months.

Water demand is generated through irrigated agriculture, municipal water supply demands, and industrial needs. Agriculture is the main consumer of water at certain times of the year, particularly at the end of the growing season, for wheat irrigation and just before the monsoon in the early stages of rice cultivation. Municipal demand remains relatively uniform throughout the year, while industry represents a small but growing water demand. In each case, demand exceeds the consistent and reliable water supply.[9]

Studies of water quality in the Bagmati and its tributaries have been conducted since the 1970s and include a comprehensive report (His Majesty's Government of Nepal [HMG] and Asian Development Bank [ADB] 1991), which collected primary data from the Bagmati and its principal urban tributaries—the Bishnumati, Dhobi Khola, Manohara, Hanumante, and Tukucha Khola.[10] This report compared tested parameters to World Health Organization (WHO) standards. Measures of pH; total dissolved solids (TDS); dissolved oxygen (DO); biochemical oxygen demand (BOD); the organic pollutants ammonia, nitrite, chloride, and organophosphate; total coliform; and the heavy metals chromium, arsenic, and copper demonstrated intensifying pollution levels from the entry of the Bagmati into the urban reach to its exit at Chobar.[11] By their points of confluence with the Bagmati, most tributaries exceeded WHO standards for several water quality parameters.

The HMG/ADB study (1992) suggested both a spatial and temporal decline in water quality, concluding that municipal sewage is the main source of river pollution, with effects including a very low dissolved oxygen content that sometimes reached zero, high BOD, and high ammonia concentrations. Large numbers of fecal coliform bacteria were found in all types of effluent samples, including those drawn from storm sewers. Although levels varied seasonally, and to some extent across study years, they always exceeded WHO standards.[12] The study also found heavy metal concentrations to be very low, indicating that industrial contaminants were not a major problem in the river system, despite factory discharges directly into the Bagmati. Chromium contamination in some areas was identified as a cause for concern, however, particularly in the vicinity of the Bansbari Tannery on the Dhobi Khola.

An undated study conducted by R. R. Shrestha suggests that Bagmati River water quality trends are best understood by looking at water quality data collected at Sundarijal (the Bagmati headwaters) and Sundarighat (where the Bagmati exits Kathmandu). This study demonstrates that BOD and chloride ion concentrations increased more than fivefold between 1988 and 1999 and that the water quality of the river worsened considerably as it moved through the city.[13] This report also shows seasonal variations in selected water quality parameters (BOD_5 and $N\text{-}NH_3$) and seasonal fluctuations in water discharge rates. $N\text{-}NH_3$, or ammoniacal nitrogen, is a compound consisting of nitrogen (N) and nitrate (NH_3), often found in landfills. $N\text{-}NH_3$ usually leaches into groundwater, where it can cause eutrophication and an imbalance to a water system's equilibrium.

In 2002, the Community Led Environmental Action Network (2002:VI-11 to VI-14) published water quality data for the Bishnumati that resembled those of the Bagmati. This study describes the Bishnumati between Teku and Balaju Bridge as "an open sewer [that] has almost lost its assimilating and self-purification capacity" (VI-11), and it provides a set of water quality profiles to substantiate its claims.

Urban construction activity is thought to have considerably altered river morphology in the Upper Bagmati Basin over the past four decades, particularly in urban river reaches. A pervasive account of morphological change in development reports is that both the Bagmati and Bishnumati were previously braided together and had flowed over sand and gravel beds in the urban area. Cross sections are thought to have been relatively flat, whereas at present the rivers are confined to deep channels. Between Teku and Thapatali, the Bagmati riverbed is believed to be about 2.5 meters lower than what is considered to be the level of the former bed; this change is thought to have taken place over the last decade and is attributed primarily to commercial-scale sand extraction (Stanley International et al. 1994: appendix 6, 6; Shakya

2001:292). An elaborate system of ghats and temples provides some idea of former riverbank edges; at present these markers are usually a considerable distance—up to 25 meters—from current channelized river flows. As with the Bagmati, sand extraction in the Bishnumati riverbed is considered the main cause of severe morphological change (see Community Led Environmental Action Network 2002: vi-30).

The Bagmati Basin Management Strategy estimates that the rivers of the Upper Basin carry about 100,000 cubic meters of sediment per year on average, mostly during peak flood events.[14] In 1992, IUCN estimated the total annual sediment discharge at Chobar as 760,500 tons per year (Stanley International et a1.1994:23; appendix 6, 3).

Despite these high sediment loads, sand deposition does not occur at a rate comparable to commercial sand extraction, and current sand-harvesting practices are considered to be unsustainable.[15] Used primarily to create cement mortar for construction, sand is extracted at both large and small scales from riverbanks and beds. A licensing system exists, but that accounts for only a small fraction of total sand harvesting.

Most of Kathmandu's domestic wastewater that enters sewers and surface drains is untreated; this water flows directly into the Bagmati and Bishnumati rivers and their tributaries. In addition to wastewater and sewage from domestic sources, the rivers also receive inputs of untreated storm water directly from city streets.

The BBMS estimated that 400–450 tons of solid waste were produced daily in the Kathmandu Valley in the early 1990s. The government's Solid Waste Management and Mobilization Center (SWMMC) handled an estimated 60 percent of this, while much of the remaining 40 percent was dumped along riverbanks. The Community Led Environmental Action Network (2002) study of the Bishnumati Corridor reported that only half of the city's total solid waste was collected by municipal or solid waste management authorities (VI-29). The banks of the Bishnumati were particularly affected by solid waste dumping; an estimated 42,000 cubic meters of municipal solid waste had accumulated along the banks of the Bishnumati at Teku over an unspecified period (Stanley International et al. 1994: appendix 2, 8).

Industrial districts, tanneries, and carpet processing plants also contributed various pollutants to the river system in the 1990s. The previously mentioned Bansbari Tannery processed about ninety tons of hides per year, and it released chromium by-products directly into the Bagmati. The carpet industry was also an important source of industrial waste inputs.[16] The BBMS cited high levels of potassium, nitrate, ammonia, and orthophosphate in river water samples collected from small agricultural catchments as well (Stanley International et al. 1994: appendix 7, 12–13).

A biophysical degradation summary was the foundation for nearly every report I reviewed in the conservation library in Dhobighat. But it was also often supplemented with references to the rivers' historical and cultural value. The BBMS, for instance, describes a Bagmati and Bishnumati riverscape in a state of both ecological and cultural collapse. Calling the rivers the center of Kathmandu Valley cultural heritage, the study embeds the scientific contours of river degradation in a cultural degradation narrative, alternating between natural and cultural formulations of degradation. This acknowledgment of a cultural context for the science gave the official restoration agenda of this period an important sense of moral urgency, adding to the technical aspects of sewage and sand extraction an appeal to the unique cultural character of the Kathmandu riverscape. Historical riverbank ritual practices are described in this way in the BBMS: "Restoration of culturally important monuments is both difficult and extremely costly and there are only a few places where such work is being done . . . all with foreign assistance. However the importance of temples is directly linked to the natural environment and purity of river water and many people feel that the pollution problems need to be addressed before or in tandem with temple restoration" (Stanley International et al. 1994: annex 3, 11).

The BBMS describes a "pre-development" (pre-1951) river history in which Kathmandu is said to have hosted a unique, ecologically balanced ritual culture. Emphasizing the rivers' traditional significance as centers of landmark Hindu life-cycle rituals, the study explains Valley Hindus' regard for water as curative, health-giving *jal*; describes practices like *śraddhā*, a series of rites following a funeral; and the use of structures like carved stone *arghajals*, where the dying could be placed so that their feet could touch flowing river water. Contemporary changes in ritual practice on the river are then linked to ecological deterioration, implying a somewhat direct cause-and-effect relationship. In passages like this one, the discussion essentially ignores the possibility that other factors might also lead to changes in Kathmanduites' religious practices or the frequency with which they used the rivers for ritual purposes; instead it assumes that degradation causes this.

> Now very few ailing people are brought to the river.... The water is no longer potable.... Many of the *sattal*s are in ruin.... many people have abandoned the ritual of dipping the feet of the dying into the water.... In really polluted areas nearby taps are used for water that's poured into the mouths of the dead.... Funeral parties are starting to bathe at the taps and just sprinkle a bit of river water over their heads. Ashes are still placed in the river but the thin trickle of water that runs through mounds of garbage and building debris does not any longer seem to possess the power to send the soul to heaven. (Stanley International et al. 1994: annex 3, 22)

The BBMS portrays the rivers of the past as being environmentally healthier and, as an implied result, culturally more "intact." In the past, it reads, "Nobody washed their clothes near the sacred ghats but would carry water home for use. The regulation of domestic affairs also existed in urban areas where it was forbidden to throw garbage in the streets except late at night and early in the morning, when sweepers would dispose of it before the business of the day began" (Stanley International et a1.1994: annex 3, 20).

The question of whether Kathmandu has historically been ordered and clean or disordered and dirty has been roundly challenged (see Tuladhar 1996:366–67), but perhaps more importantly, by using a particular and selective rendering of cultural history that identifies specific cultural practices as central to ecological stability, the BBMS gives its biophysical portrait of degradation a sense of moral urgency. Where BOD and coliform statistics define an urgent water quality problem, the withering of particular and unique cultural practices conveys a relationship between environmental and cultural loss. References to the riparian templescape and the historically sacred quality of the Bagmati thus mobilized affective sentiment on behalf of the modern, developmentalist ideas of urban ecological progress that reports like the BBMS ultimately espoused. In a sense, this took ideas and practices more conventionally considered premodern, or contrary to modernization, and put them to work in the service of developmentalism. The public legitimacy of attendant development projects, after all, depended not only on the soundness of the science but also on moral authority.

Although largely unelaborated in official reports of river change, references to the riverscape's sacred value nest into a deep and complex historical relationship between social practices and water sources throughout the Kathmandu Valley. The precise form of this relationship is actively debated and is the subject of a robust literature. To return to a scholar whose work on Kathmandu's sacred landscape was mentioned previously, Mary Slusser has noted, "As social centers and, in effect, hypaethral shrines, the water sources [of the Kathmandu Valley] serve human needs far beyond the mere provision of water. Water itself is sacred, as is everything that relates to it—the vessel, the well, the pond that contains it, the fountain from which it issues, or the stream in which it flows. Providing access to water is thought to be especially meritorious . . . thus century after century of construction by king and commoner . . . has left no corner of the Valley without a liberal number of water sources" (Slusser 1982:154).

Among water sources, rivers assume foremost importance; their water was historically used for domestic and religious purposes, and it has for centuries been infused with powerful symbolism derived largely from Hindu cosmology. In this sense, BBMS references to jal, śraddhā, and arghajals are an entirely appropriate starting point for an inquiry into the ways one might assess intersections between river degradation and cultural change.

Sites along the Bagmati have historically served as centers of Hindu ritual, where the river's divine power may be invoked to purify, absolve sin, and rejuvenate (see also Alley's discussion of the somewhat analogous power of Ganga [2002:55]). Historically, the Bagmati was revered through worship (*puja*), oil lamp rituals (*aaraati*), and ritual ablutions (*snaan*) in the river. The ashes of the cremated were immersed there to ensure safe passage to an ancestral realm.

The contemporary complexity of concepts and practices related to a sacred river is brilliantly engaged by Kelly Alley in her recent study of material and ritual pollution of the Ganges in India. In a thorough ethnographic portrait of an Indian Hindu religious frame for river ecology, she shows how overlapping semantic domains of Ganga and *gandagi* (a common term referring to the material conditions of waste) produce conflicting ideas about water pollution.[17] Alley shows how "Hindus in Banaras subordinate the destructive power of *gandagi* to a sacred power that creates its own ecology.... In their view, this sacredness transcends people and the powers of the natural ecologies they construct" (2002:78, 79). Conditions that a global environmentalist audience might automatically classify as pollution, then, are regarded among many of Alley's devout informants as ultimately dissolved and overpowered by the Ganga (e.g., 80), fixing, in important ways, river sacredness and reinforcing its timelessness (230), but also positioning it as beyond human scientific or materialist power. She goes on to observe that "there is no separate domain for 'nature' in the Hindu discourse. Instead, Ganga, a deity, subjects (in very humble and loving ways) both humans and ecological processes to her own sacred design" (2002:238).

A study of the Ganges is extremely relevant here since the Bagmati eventually joins this river, which is considered most sacred among Hindus. The Bagmati is itself considered to be second in sacred significance only to the Ganges, and since the Ganges flows entirely within India, it is the Bagmati that is the most sacred river in Nepal for Hindus.

A rich body of scholarship preceded and follows Alley's important work on South Asian Hindu religious belief and ecology, with important attention to water and rivers specifically;[18] recent attention has produced rich analyses of the interplay between scientific fact, nationalist narrative, and Hindu identity formation vis-à-vis sacred spaces and the environment in India (e.g., Mawdsley 2005, 2006; van der Veer 1994).

In the current case, however, encounters between a contemporary, officially Hindu Nepali state, its development apparatus, and river-concerned activists are our focus, and among these actors, the material conditions of waste were not directly contested.[19] That is, the sacred power of the Bagmati to overcome or retain its sacred quality despite material inputs was not the explicit center of river activist and official concerns. Instead, invocations of a sacred Bagmati were used selectively and strategically to promote competing ideas about how to achieve a modern city with a modern, restored environment while retaining or reclaiming the identity-making power of a riverscape with deep cultural history. The place of Hindu ritual practice in that restoration was central only for some in the present study, and even they did not claim that centrality as an automatic remedy for material pollution. So as Alley's work illuminates how wastewater was interpreted when it met the sacred Ganges, this case notices how reverence for the sacred figured directly in visions of a capital city's modern riverscape and future.

The BBMS—a development policy document first and foremost—endorsed restoration goals that included templescape restoration and the construction of riverside parks and open spaces. With specific regard to river flow, the report called for the construction of an extensive system of weir dams that could capture and replenish sand during high sediment-load periods, thereby restoring the rivers' previous morphology. A raised riverbed was expected, in turn, to raise and widen river flow, so that river edges would once again touch the stone ghats and walkways lining the urban templescape. The need to treat surface water inputs and to enact and enforce river management regulations were noted. But an important element of the transition to a restored riverscape was simply missing: how would a highly degraded riparian zone almost completely occupied by informal settlements (that is, the homes of a large population of landless poor) be transformed into the clean, green, open, public parkland that the report envisioned? One could only pose this question and infer that what was left unanswered was somehow not central to urban ecology or urban river restoration or, perhaps more likely, it was not politically possible to find an answer.

Official framings of river degradation in reports like these constructed an authoritative, biophysical account of the problem for those with the presumed power to solve it, Nepal's government and development agents. Meeting particular biophysical objectives, these reports suggested, would enable continuities in cultural practice and meaning between Kathmandu's past and its future.

Missing from this framing of river decline and revival was a clear and reflexive acknowledgment that each proposed project was tied to, and reproduced, the web of development connections on which the Nepali state had grown to depend for its own survival. Also strikingly absent was a clear and transparent plan for the future of the tens of thousands of landless migrants that by the late 1990s were settled along the riverscape. Implicit, perhaps, in this omission was that as long as the power to effect change remained in the hands of the state and its development collaborators, the population of migrants—outsiders to legitimate river ecology according to this frame—would have to be moved, whether or not it was consistent with their will. In this sense, official urban ecology enabled the exclusion of a particular category of people, while at the same time it reinforced the state-development complex as the appropriate authority to save, protect, and conserve the Bagmati and Bishnumati rivers into the future.

A HOUSING ADVOCATE'S URBAN ECOLOGY

By combining their own data with a sense of moral urgency, housing advocates argued against the automatic and total exclusion of migrants from a more ecologically sound riverscape. Responding to environmental decline in the city as well as new patterns of rural-to-urban migration, these advocates defined urban ecology in terms quite different from the official narrative. Even as they drew on international development institutions for discursive strategies to express their position, they sought to empower those most marginalized by development and modernization processes. In a housing advocacy framing of urban ecology, the crisis was one of inadequate shelter first, with environmental stress as its dire consequence. Moral traction derived from a logic of universal human rights, most specifically a universal right to adequate shelter.

The crisis to which housing advocates were responding was in full force through the 1990s, as Nepal experienced growing rural-to-urban migration. Between 1991 and 2000, the portion of Nepal's population residing in urban areas increased from 9 to 15 percent. In 2001, the noted scholar Sudarshan Raj Tiwari suggested that this indicated a major shift in Nepal's migration trends, from historically rural-to-rural and interregional migration patterns to permanent rural-to-urban mi-

Figure 1. The Bagmati River just below the Bagmati Bridge, 2006. A tiered platform, with a fish ladder at the lower right, supports the bridge.
(Photo by the author)

gration. Tiwari attributed the shift to deepening rural poverty and what he called the "saturation" of rural agricultural land:

> Although initial urban settlements started as a result of pull factors generated by the development of access and concentration of administrative, trading, manufacturing and other economic activities, their growth, in many cases, is spurred by rural-urban migration resulting out of impoverishment of rural areas and its inability to sustain the growing population. Poor agricultural productivity and segmentation of agricultural land has been pushing rural people out to urban areas. In a way, this process of urbanization has transferred rural poverty to urban areas and many urban fringes show rural characteristics. (2001:2)

These "rural characteristics," if they are to be understood in Tiwari's socioeconomic terms as evidence of an economic periphery physically present in the economic center of Kathmandu, were increasingly visible on the banks of the Bagmati and Bishnumati rivers throughout the 1990s. During this period, as noted earlier in the book, growth in Kathmandu catalyzed the rapid spread of urban development over a large area and increased population density throughout the city.[20] Despite a construction boom, housing supplies were insufficient to meet overwhelming demands, and rent prices became prohibitive. For new migrants and poorer city residents, participation in runaway land and housing markets was impossible, and the exposed sandy beds of Kathmandu's rivers became a logical settlement destination.[21] To some extent, the rivers increasingly displayed the human consequences of decades of uneven development and rural underdevelopment, and those consequences were brought evermore proximate to Nepal's political and administrative elites.

Tiwari's designation of a turning point in migration patterns highlighted how rural poverty, in newly visible urban concentrations, entered popular urban consciousness in new ways.[22] "Rural characteristics" on the urban margins defined a population that the city resisted by retaining poor migrants' "rural-ness"; for Tiwari they represented the failure of the very development processes that the city had historically promoted.

Perceptions of a connection between rural-to-urban migration and urban river degradation were fueled by their apparent simultaneity during the 1990s. In 1991, riparian *sukumbasi* settlements were estimated to be growing at 12 percent annually, a rate twice that of the city itself (HMG/ADB 1991). In 2001, the growth in the number of squatter settlements continued at a rate of 12–13 percent. Between 1990 and 2000, the total number of urban squatter settlements almost tripled, with a majority located on public lands along rivers (Hada 2001:154).[23] By 2002, a significant percentage of Kathmandu's riparian corridor was lined with semipermanent sukumbasi housing.

There is no comprehensive translation from Nepali to English for *sukumbāsī*, although it is most commonly translated to "squatter." A related word, *sukumbāsa*, is the state of having nothing. Used to refer both to people and their settlements, sukumbasi refers to those who are assumed to be landless, or very poor, and who occupy land for which they do not own a legal title. Although technically the term refers to "the person lacking shelter and food; one having neither" according to Pradhan's (2001) *Ratna's Nepali English Dictionary*, the Bagmati and Bishnumati sukumbasi populations have become "permanent" parts of the riparian landscape such that they cannot be said to be definitely and universally lacking food and shelter. The population is made up of both rural-to-urban migrants and migrants originating within the Valley.[24] Middle- and upper-class residents of Kathmandu tend to refer to anyone who occupies informal housing on public land as sukumbasi. It should be recognized that the term can carry negative connotations and, although it is widely used, can be taken as an insult.

A perceived inundation by rural-to-urban migrants weighed not only on river debates, but also on broader discussions of the general decline in Kathmandu's environmental quality during the 1990s. Like many others, Tiwari's migration analysis goes on to link rapid rural-to-urban migration to the variety of urban social and environmental stresses that had, by the turn of the twenty-first century, come to characterize contemporary Kathmandu: "The visible consequences of such urban growth have been the depletion of natural resources, failure of basic urban services to support the town, environmental deterioration, lack of housing, and health problems of increasing magnitude" (2001:3).

And yet, despite strong discursive characterizations of riparian settlers as features, and sometimes agents, of ecological and cultural degradation, little official action was taken to manage migration. At the level of city government,[25] the question of rural-to-urban migrants received increasing rhetorical attention, but few policies were actually implemented to regulate settlement growth. Writing for a development policymaker audience, Hada (2001) observed a stark incongruity between the content of river management planning documents and actual implementation:

> There are no clear cut policies related to city squatters in Kathmandu. Although the National Plan of Action [stated an intention to] upgrade and manage unplanned settlements, the details of [this plan] have never been worked out. The Concept Plan for Bishnumati Corridor by Halcrow Fox Associates, 1991, . . . proposed [the] relocation of existing squatters along the Bishnumati riverbed affected by the proposed development of the Bishnumati Link Road. The proposed relocation scheme included a low-income housing project in another site nearby which should be supported by closely related programs such as long-term housing loans, appropriate savings

schemes, and lending parameters specially targeting the poor or low income households to invest in income generating activities as well. (160)

Not only was the Bishnumati Corridor squatter relocation scheme left dormant until much later (see chapter 6), the Bishnumati Corridor Environmental Improvement Project actually resulted first in forced squatter evictions, not relocations.

This disconnect between planning documents and actual practices was often attributed to an unstable set of bureaucratic institutions under democracy. By 1999, for example, the municipal Commission for the Resolution of the Landless Squatters Problem had seen twelve committees form and dissolve over ten years, one for each new government that was elected and ousted.[26]

The increasingly visible and seemingly simultaneous growth of riparian settlement on one hand, and intensifying urban river degradation on the other, facilitated a popular urban belief that there was a causal link between the two. This link, in turn, obscured the complex array of factors that contributed to urban river degradation, foregrounding instead the perceived destructiveness and invasiveness of informal river-proximate populations. Ecological narratives became a way to simplify the equally complex problems of accelerated rural-to-urban migration and intensified river degradation.

Through conferences, media coverage, attempts to raise public awareness, and official negotiations, housing activists countered dominant perceptions that sukumbasi were an obstacle to river restoration through an alternative, albeit ecologically reasoned, narrative of ecology. By emphasizing concepts like "healthy cities" and "sustainable human settlements," ideas associated with the United Nations Habitat Program,[27] housing advocates foregrounded socioeconomic concerns in discussions of river ecology. Framing their vision of a restored riverscape in terms of a sustainable urban future, they called for upgrading sukumbasi settlements by improving public health, education, and sanitation conditions within them. These goals were portrayed as fundamental aspects of urban ecology.[28]

Although they did not directly contest the idea that the Bagmati and Bishnumati rivers were degraded, housing advocates and sukumbasi activists argued for a completely different fate for sukumbasi in their framing of restoration discussions. Upgrading settlements rather than removing them was regarded as the key to realizing an ecologically healthy riverscape. Claims to culture and history were in fact generally missing from the sukumbasi perspective; instead, a focus on urban modernity and the urban future through appeals to notions like forging a "sustainable habitat" dominated housing advocates' approach to ecology.

Kathmandu's leading sukumbasi advocacy organization, Lumanti, hosted the Future Cities World Habitat Day Conference in Kathmandu in 1997 (e.g., Bajracharya and Manadhar 1997). The conference served as a local follow-up to the UN Conference on Human Settlements held in Istanbul in June 1996. Throughout the session, phrases and terms linked directly to that conference, particularly "sustainable human settlements" and "habitat," were localized by their application to the plight of Kathmandu's sukumbasi. Never did these ecological terms imply a threat to sukumbasi or their place on the city riverscape, however. Rather than blaming a riparian human presence for river pollution, for example, panelists pointed to insufficient infrastructure. Rather than regarding riparian sukumbasi as a disproportionate cause of river degradation, they were described as the disproportionate sufferers of the consequences of degradation. As city residents most proximate to the rivers, river pollution and degradation were regarded as immediate and serious threats to sukumbasi themselves, rather than the other way around.

Missing from housing advocates' framing of ecological restoration were pleas for open park spaces or the renewal of dying cultural traditions. Instead, "habitat" restoration carried connotations of contemporary human rights and urban infrastructural development. Conference panels and proceedings emphasized the idea that responsible urban environmental management called for the creation of "sustainable human settlements" in which ecology signaled needs like human housing, food, and shelter.

In interviews, housing advocates consistently resisted any suggestion that sukumbasi settlements had adverse effects on the ecological integrity of the rivers. Proximity to the degraded resource *seemed* to implicate sukumbasi in the degradation process, they agreed, but in realistic ecological terms, sukumbasi actually played a relatively minute role in degradation. This was most obvious in the case

Figure 2. Permanent sukumbasi housing at Shankhamul, 2006.
(Photo by the author)

of sewage, which the BBMS identified as the single most important contributor to water quality deterioration in the urban system. Since Kathmandu lacks a comprehensive, functional sewage treatment system, effluent inputs originate throughout the city, implicating "legal" and "illegal" urban inhabitants alike. Housing advocates advanced this point by asserting that since sukumbasi consumed relatively fewer consumer goods than more wealthy urban inhabitants, their likely per capita contribution of chemical inputs, solid wastes, and other by-products of industrial production to the river system was also relatively small.

Sukumbasi and their advocates also argued that the positive impacts that migrants had on the river system—which in real ecological terms might have been minor but could nevertheless be regarded as evidence of awareness and action on behalf of the rivers' condition—were completely overlooked by state-development bodies (e.g., Lumanti Support Group for Shelter 1998:4–5). At the level of individual and collective micro-management, they argued, many sukumbasi communities were actively engaged with the rivers, performing tasks, either collectively or individually, that constituted efforts at river quality improvement.[29] For example, housing advocates pointed out that planting vegetation on the riverbanks—from trees to agricultural crops—was a common migrant practice, while perhaps ironically, riparian re-vegetation was a goal of several state-development interventions.

By claiming rhetorical tools from the international development arena, housing advocates countered the state-development narrative on the development industry's own terms, redefining urban ecology in a way that served riparian squatters' eco-political objectives.[30] By emphasizing a housing and growth crisis in the present and by avoiding an either/or positing of cultural heritage and riparian settlements, this version of river degradation and restoration avoided historical portrayals that would locate sukumbasi outside the legitimate river system.

This is not to suggest that river water quality was unimportant among housing advocates. But in emphasizing technical changes that were needed, such as developing sewage treatment, rather than looking at reclamations of history and cultural meaning, housing advocates avoided a narrative that might deem sukumbasi as invasive. This framing also tended to de-emphasize the vulnerability of riparian settlements in general, however. Since monsoon flooding can be swift and dangerous, this often left those fighting against eviction in the name of ecosystem improvement as appearing to fight for settlement sites that left migrants vulnerable to flash flooding. In addition, legitimizing existing riparian sukumbasi settlements essentially eliminated the possibility that past river flow levels and riverbank contours could be restored, since by the late 1990s a majority of settlements were located in former riverbeds.[31] If existing riparian settlements were legalized and upgraded, then a certain idea of ecological improvement might be realized, but the state-development vision of widened river flows near ancient ghats, central to its idea of river cultural heritage, would be impossible.

In contrast to a state-development frame for urban ecology, housing advocates de-emphasized cultural history, choosing instead to concentrate on the modern present and its dire problems. Visions of restoration started with present conditions and proceeded toward a sustainable urban future. That future required its adherents to background some biophysical and historical details and emphasize instead the moral urgency of a crisis of inadequate urban shelter. The riparian zone was thus framed as human habitat first; its human population faced the question of *how* to stay in a more sustainable way rather than *whether* to stay at all.

Housing advocates, organized as nongovernmental entities and therefore themselves dependent on foreign donors, did not categorically oppose political collaboration with Nepal's government and international development interests. However, their overall agenda for the empowerment of migrants prioritized political inclusion for some of Kathmandu's most marginalized citizens. Promoting a sustainable city thus involved imagining a fully participatory democracy that would lend this population an active voice, an aspiration throughout the 1990s.

A CULTURAL RESTORATIONIST'S URBAN ECOLOGY

A third frame for knowing the problem, which I have labeled a cultural restorationist view, rejected official and housing activist approaches. Its diagnosis of river degradation foregrounded the role of international development agents and government, which it charged with failure to protect the city's environment, and, in turn, the Valley's cultural integrity. To explore a cultural restorationist's narrative, we return to one of its prominent figures, Huta Ram Baidya, who made the following declaration in the opening passages of this book: "Development has destroyed my cultural rights!" For Baidya and his collaborators, the very idea that the modern state-development complex could restore and protect the cultural integrity of the riverscape had no basis in fact or experience.[32] For him, the dominant frame through which the rivers should be understood required a complete refocusing through the lens of what he called the Bagmati Civilization (*bagmati sabhyata*).[33]

During the 1990s, Baidya's views circulated in public media, from Nepali- and English-language newspapers to radio, television, and Internet web sites; through these, he emerged as a spokesperson for a predominantly elite and middle-class heritage-focused conceptualization of environmental change. As a counternarrative to government and housing advocate frames of the problem, the Bagmati Civilization narrative described an ideal moral geography that was at once socially exclusive and politically defiant.[34]

At the center of the Bagmati Civilization narrative was the argument that the rivers' demise was integrally connected to a collective forgetting of the Kathmandu Valley's cultural and natural history. Only through "remembering" that history could the ecological integrity of the rivers be reclaimed; a more desirable, ordered polity was expected to follow (Rademacher 2009).

In one of our earliest encounters, in October 1997, Baidya proclaimed that he was on a "one-man crusade to save Bagmati Civilization." Over time I learned that his crusade involved an approach to river restoration and protection that embraced his idea of an ecocultural whole that included his account of the rivers' present status and his version of the myth, prehistory, and history of an interconnected riverscape and Valley polity. Baidya's description of Bagmati Civilization history identifies two consequential strains of forgetting: a general social forgetting of the rivers and their significance and an administrative forgetting—a failure of the state to adhere to its river stewardship responsibilities. It is this latter form of neglect that animated his Bagmati Civilization idea with a deeply nativist antidevelopmentalism.

Baidya was outspoken in his criticism of the relationship between the state, development industry, and the cultural dysfunctionality that he associated with river degradation. He emphasized Kathmandu's haphazard contemporary history of urban planning in general, and planning on the rivers specifically, to directly link the dependent relationship between the Nepali state and foreign aid to river degradation.[35] Baidya argued that true reclamation of the river system was possible only through autonomy from international development donors and with the creation of an accountable, and what he considered truly democratic, central government.

To demonstrate how development itself was a central culprit in urban environmental decay, Baidya regularly accompanied anyone who expressed interest in a walking tour of specific river projects and infrastructure. Included on the tour were bridges that were built during the last decade, most of which were engineered according to the degraded riverbed morphology at the time of construction. If river restoration was predicated on a basic need to repair and raise the deeply channelized riverbed, thereby raising river flow levels, then each new bridge constructed to degraded riverbed specifications appeared to Baidya to further limit the likelihood of restoration itself. Therefore, even as major river management plans like the Bagmati Basin Management Strategy identified raising the riverbeds and widening river flow as fundamental goals of restoration, actual official practices made achieving these goals both difficult and unlikely. Baidya's point was that new state-approved and development-sponsored bridge construction undermined the feasibility of officially stated restoration visions, either demonstrating an extreme lack of coordination or bringing into question true state-development motives. He argued that such infrastructure projects spoke louder than the commitments to cultural heritage or river system improvement that were written into planning documents or espoused in official rhetoric.

Baidya also criticized the territoriality of development, noting that donor relationships seemed to allot particular development objectives—or even parts of the city—to certain donors or countries. In the late 1990s, he cynically referred to the Patan Bridge area as "Tokyo Plaza," since much of it had been built up or restored with Japanese development assistance. He was particularly frustrated with the way that the Gopal Mandir was marked after a partial restoration: a plaque was affixed in a prominent temple entranceway declaring that it had been "built through the joint efforts of the Nepal Association for the welfare of the Blind, the Tokyo Helen Keller Association with funds provided by Japan Ministry of Postal Services, Voluntary Deposit for International Aid." After a few weeks, someone scrawled a correction in black marker on the white plaque, making it read, "RE-built through the joint efforts." The plaque itself remained in place, however, and was displayed far more prominently than an original inscription identifying the temple's patrons from the eighteenth century—who, while also motivated by political objectives, were considered by Baidya as more authentic and historically accurate "builders." For Baidya the plaque was clear evidence that in both their conduct and in the forms

Figure 3. Huta Ram Baidya records river conditions during a river walk in 2000. *(Photo by the author)*

they created, international development agents exercised far too much control in the capital.

Examples abounded; all substantiated Baidya's charge that international development projects were ecologically and culturally destructive. On several occasions, Baidya accompanied me to the site where two ghats were destroyed to create space for the new Bagmati Bridge in 1997, and we also went to the place where a former monastery on the bridge's north end had been converted into a children's park. Along the Bishnumati, he regularly pointed out the enormous public toilet buildings erected on the riverbanks as part of the Bishnumati Corridor Environmental Improvement Project. One such facility, which, like the others, drained untreated wastes directly to the river behind it, was built just a few feet from a frequented temple site. Such development measures were, for Baidya, clear violations of the rivers' cultural value.

Baidya viewed river degradation as a cultural problem inseparable from questions of political control. Nepal's state-development alliance produced an uneven, distorted relationship between international aid, its experts and planners, and the Nepalese public, he argued; this relationship was, in turn, a primary factor in river degradation. Like many others active in river politics, Baidya referred to a connection between democracy and accelerated environmental decay over the 1990s, but he reasoned that what was referred to and supported by state-development interests as "democracy" was a misnomer. In reality, he argued, development and environmental policy were profoundly undemocratic, largely dominated by international donors. Baidya pointed to blatant contradictions between what state-development policy documents said, and what was done in practice, ultimately concluding that the relationship between the state and international development institutions was itself deeply flawed. The state in this relationship was not only undemocratic but perilously weak, evidenced by its failure — among other things — to enforce existing limits on the extraction of sand, which contributed to the collapse of several bridges. Accordingly, river restoration would require state autonomy from donor influences and foreign experts. That autonomy would have to be demanded before it would be granted, making the cultural heritage restorationist's ecology also an urgent call to political action.

Such action could only be taken by a populace fully conscious of the unique cultural attributes of the Kathmandu Valley and its rivers. Baidya referred to this consciousness as "remembering the Bagmati Civilization," and he elaborated the details of Bagmati Civilization history in widely circulated publications, interviews, and speaking engagements. I elaborate the layers of the concept elsewhere (Rademacher 2008; 2009), but consider just one early example of Baidya's description of a riverscape lost to degradation. In his essay "The Holy River Bagmati," which appeared in a preservationist collection of photographs and essays published in 1995 called *Images of a Century*, he instructs his largely elite, international-development community readership about the basics of the "Bagmati Civilization."[36]

Readers learn that the Bagmati was historically a crucial historical site of religious training, discipline, and practice. Strict adherence to rituals of homage, purification, and the reproduction of holy sites through pilgrimage once formed a social order, continuity, and collective respect for the river. These, in turn, produced river cleanliness and ecology, creating a nature-culture functionalism essential to the river's past vitality.

Historical riverscape practices are said to have created social equity, despite an otherwise rigid social hierarchy: "People of all castes and social status," he writes, "have contributed to protect, decorate, and structurally stabilize *tirthas* and riverbanks throughout the centuries." He recounts the temples and monasteries that modern kings and Rana leaders built along the river, deeming all to be culturally important regardless of the terms of each individual's rule. Even the infamous acts of Rana rulers like Jung Bahadur Rana could be "washed away," Baidya suggests, in the act of building up the riparian templescape.[37]

The essay also highlights Baidya's personal memories, which combine old devotional practices with a powerful sense of social obligation to the riverscape. "Mindful of public warnings," he writes of the past, "no one dared pollute the river and its surroundings." The reader is encouraged to imagine with Baidya a festive past scene of Shiva Ratri, in which he walks along a "clean, stone-paved ghat," lined by rows of willows. The implicit unity of natural order and specific cultural practices forms a key foundation of the Bagmati Civilization narrative.

In essence, "The Holy River Bagmati" traces river degradation to what Baidya calls forgetting: forgetting the river's divine origins, forgetting the social order of past ecological balance and regulation, and forgetting the government's "social and administrative responsibility" toward the rivers. Baidya's characterization of the present emphasizes inundation: a chaotic, unplanned influx of outsiders, the consequent overdrawing of water and accumulation of waste, and the general disaster of Valley development. The development industry, in alliance with the state, as well as the ideas of progress with which development processes are associated, are said to have exacerbated riverscape devastation in the present.

The charge of forgotten history is not simply left to the imagination in *Images of a Century*. It is supplemented by a set of photos that include a monsoon-swollen Bagmati flanked by swimmers or worshippers, photographed in the 1920s, and the wide, shallow Bagmati of the 1940s, braided with ripples of sand.

The emphasis for a cultural heritage restoration frame of urban ecology, then, was the destructive interface between modernization and development on one hand, and a particular rendering of cultural history on the other. By portraying that cultural history as lost to invasive and undemocratic modernity, Baidya equated river restoration with the very reclamation of cultural identity and political autonomy. The emphases were not explicitly biophysical; they were fueled instead by a moral urgency to recover the political capacity to create and re-create the practices that made Baidya's Bagmati Civilization meaningful. The outsiders in this frame were international development interests at one end, and culturally "forgetful," or culturally different, Nepali citizens on the other. This latter point is critical, as an implicit effect of the Bagmati Civilization narrative is a clear accounting of which identity, class, and caste groups could stake legitimate claims not only to the riverscape but to the Kathmandu Valley itself. If the crisis of the rivers was a crisis of forgotten history, only those with a legitimate claim to that history could undertake effective restoration. This excluded foreign development entities, and to a large extent, those citizens of Nepal and Kathmandu who failed to meet Baidya's fairly complex criteria for belonging.

Whereas inadequate shelter stood as evidence of state-development failure for housing advocates, the crumbling templescape was evidence for Baidya. But international legitimacy for the Bagmati Civilization — discursive or otherwise — was not directly sought; in fact, the moral power of Baidya's narrative hung in part on its critical distance from the development sphere. The carved stone arghajal that gave evidence of Baidya's loss of "cultural rights" in the opening passages of this book, then, was framed as a political and development disgrace.

KNOWING THE PROBLEMS IN TIME AND PLACE

These contrasting ways of engaging and defining river degradation may demonstrate the malleability of what urban ecology "means," but their simultaneous existence is analytically unsurprising (e.g., Blaikie 1985; Blaikie and Brookfield 1987; Peet and Watts 1996:38). What is important to notice are the ways that each framing established in competing ways its legitimacy, its urgency, and its knowledge priorities. All of those who are active on behalf of river improvement drew on visions of precisely what that improvement looked like, and its social and political dimensions varied widely across different groups. As a consequence, an otherwise sizeable, informed, well-funded, and deeply committed collective of river-concerned activists was differentiated, and in a sense disempowered, by the complexity of the problem itself. Quite contrary to the assertion that "knowing the problem is *not* the problem," in important ways knowing the multiple forms and cultural and political stakes of the problem was *precisely* a problem. It provides a partial explanation for how billions could be spent on a river that was "still polluted."

Questions remain. How should each aspect of degradation be defined and weighted? How should society value an urban river in a modern, developing city? How should a historical and sacred legacy be incorporated into the modern present? To what extent was the contemporary Bagmati a sacred landscape, and how important was its religious value to conservation objectives? Or, as Baidya might have it, was conservation itself a path to "remembering" the sacredness of the river? Perhaps most important, who was ultimately empowered to define and effect river improvements?

It is this last question, of political transformation and its relationship to river debates, to which I now turn.

3
War, Emergency, and an Unsettled City

AS ACTORS CONCERNED WITH river restoration engaged with democratic reforms and cultural transformation, the social and environmental stakes of their actions were shaped in part by an extremely dynamic and uncertain political terrain. Although the stacks of "good reports" stored at the conservation library in Dhobighat had already suggested it, the imperative of considering interactive relationships between political change and urban ecology crystallized on a day when I was neither in Kathmandu nor engaged in formal fieldwork.

In the summer of 2001, I spent several months in Ithaca, New York, completing a summer semester of Nepali-language coursework. On the morning of June 2, a colleague greeted me by asking if I had seen the newspaper. He quickly produced a copy of the *New York Times*, and I stared at the front page, stunned. An unbelievable headline explained why I was seeing color photos of Nepal's King Birendra Bir Bikram Shah and Queen Aishwarya on the front page of this American newspaper: "Royal Family of Nepal Is Shot Dead in Palace: Prince Reportedly Opens Fire During Dinner."[1]

The article explained that the king and queen and several royal family members were killed during a dinner at Narayanhiti Palace.[2] Several others were severely injured. Crown Prince Dipendra was the suspected lone assailant, but, upon the death of his father, Dipendra had also been named the new king. Having allegedly shot himself after committing the murders, Dipendra laid in a coma in Chauni Military Hospital, where his parents were pronounced dead that same night.

International news wires ferried the story around the world, but inside of Nepal a government news blackout held citizens in suspense and disbelief. For fifteen hours, silence across all official networks produced an eerily quiet local setting in a news-saturated global mediascape. State radio confirmation that King Birendra was dead finally came from Keshar Jung Rayamajhi, the chairman of the State Council. He then pronounced Crown Prince Dipendra the new king. Given the prince's comatose condition, however, Gyanendra Bir Bikram Shah, Birendra's only surviving brother, was declared as regent. The announcement made no mention that Dipendra was the primary suspect in the killings.

For days to follow, Nepali state television and radio played only Hindustani classical music. Other news media remained shut down. In the silence, a vast chasm opened between stories circulating in the international press and those suggested or confirmed by Nepal's official news sources. The *New York Times*, identifying its sources as "anonymous government and military officials," instantly attributed Dipendra's shooting spree to an argument between the prince and his parents over the woman he wished to marry. Dipendra was said to have fought with his parents, left the room, and returned dressed in army fatigues to fire his automatic weapon on those gathered for the dinner. The BBC and CNN apparently knew of the incident before the country's own prime minister, giving foreign journalists the appearance of a curious proximity to information about the incident.[3]

Eventually, on state radio, Regent Prince Gyanendra officially confirmed that the king and queen had died in the shooting at Narayanhiti Palace. Rather than call the incident a mass murder, however, he explained it as an "accidental firing from an automatic weapon." As I sent messages back and forth with my friends and host family in Kathmandu, one friend shared her frustration at this phrase and about the general lack of official information that it exemplified. "How could it be that the rest of the world has access to an explanation while Nepal's own people are offered nothing but the preposterous suggestion of an accident?" she asked. Too many questions were left unanswered, with no indication of whether or when the government would address them.

Only a few royal family members were unhurt in the incident; notable among them was King Birendra's brother, Gyanendra, who was away from Kathmandu and therefore not at the palace on the night of the massacre. Gyanendra's wife, Komal, and son, Paras, were both in attendance, yet neither was killed in the otherwise lethal shooting. This one family's miraculous survival, combined with a history of political differences between King Birendra and his brother's family, fueled confusion and intense suspicion.⁴

The next day, June 2, brought the Bagmati riverscape into the unfolding story when the bodies of the dead were cremated at Arya Ghat, Pashupatinath. Since King Birendra's son, Dipendra, was in the hospital in critical condition, the king's brother, the Regent Prince Gyanendra, attended the funeral pyre in his stead. Sama's hope, expressed just a few months earlier, that the new king would be inspired with an enlightened sense of responsibility for river restoration on the occasion of his father's cremation, was impossible under these circumstances.

Recounting the state television broadcast of the cremations, Manjushree Thapa, a novelist and public intellectual, noted the place of the river, however degraded, in the death ritual. In *Forget Kathmandu: An Elegy for Democracy*, she writes, "At the cremation ghats by the Bagmati River lay four wooden pyres. The army conducted a salute as the brawny Brahmins carried the royal family's bodies around each pyre three times before setting them down.... The astrological chart of each person was torn and scattered into the foul and polluted, yet sacred, Bagmati River. The banks of the river had been shored up so that there would be enough water to wash these scraps away" (2005:18). The legacy of a powerful connection between the Bagmati and monarchical power, as well as the river's contemporary polluted state, were thus unmistakable frames for this moment of symbolic state transition. However ecologically degraded, the river was nevertheless host to the last life-cycle ritual of the nation-state's most powerful figure; in this, the rivers' own political and symbolic powers were also publicly reproduced.

Back in New York, I anxiously followed events over the Internet, puzzling over correspondence from friends and host family members. All avoided detail in their many messages, at once helping me to make sense of the tragedy and underlining their confusion and suspicions. Some encouraged me to reject official accounts of the incident, while others referred to Prince Dipendra's ill-fated love life or a legend that foretold the demise of the Shah Dynasty after ten generations.⁵

Months later, I returned to Kathmandu and to myriad explanations of what had happened in early June. Some insisted that Dipendra had died on the night of the shootings at Narayanhiti Palace, and that official claims that he was alive in the hospital that night were intended to dissuade suspicions that the incident was a well-planned coup—perhaps by Gyanendra, perhaps by the army or a foreign power. The quick and extensive destruction of, or failure to collect, standard forensic evidence only exacerbated suspicion. Left unanswered were fundamental questions of how, for instance, the murders had occurred and why the evidence was destroyed so quickly and completely. Citing the intensifying People's War in the countryside, some speculated that the Indian intelligence service, the Research and Analysis Wing, or RAW, was behind the massacre. Others suspected the Central Intelligence Agency in the United States. Perhaps these countries were dissatisfied with King Birendra's relatively tolerant approach to the Maoist movement; perhaps they sought a king who would exercise more overt and immediate aggression.

Confusion about the murders fueled popular speculation and rumor, and this exacerbated mounting concerns over the general disorder that enveloped the capital city, its politics, and its ecology. This disorder had been building at least since February 1996, when the Communist Party Nepal-Maoist, or *Maobadi*, declared its People's War, directly challenging state and royal authority, legitimacy, and symbolic power. Just prior to that declaration, Nepal's United People's Front, the political wing of the CPN-M, submitted a list of forty demands to the Nepali government, some of which directly addressed the monarchy. Among them were demands for a new constitution, the abolition of all privileges for the monarchy, and the elimination of royal control of the army.[6] Three days after the declaration, rebels attacked the banking system in Gorkha by burning land deeds that villagers vouched as collateral for loans. Thapa writes, "The same day, they attacked police posts in Rolpa and Rukum districts in the west, and set off a bomb at a soft-drink bottling plant in Kathmandu. They also attacked a liquor manufacturer in Gorkha and looted the house of a 'feudal usurper' in Kavre District" (2005:131).

Violent clashes between the government and Maoist rebels escalated sharply as the 1990s unfolded; the rebel army grew in size and scope while government forces employed more and more sophisticated weaponry. By the time of the royal massacre, few districts were free from violence or threats of violence. Only the capital city enjoyed relative isolation from the everyday confrontations and dangers of a nation in civil war.

The royal deaths added a new and profound layer of uncertainty to this already tumultuous atmosphere. Forging a coherent explanation for them was crucial if citizens of the Valley were to find some sense of enduring state and national order within which to situate their everyday lives. In the absence of an official narrative, speculation about who committed the murders gave way to elaborate explanations for how they could have happened. A longtime friend, for instance, talked long into the night about how multiple gunmen, each wearing identical masks of Prince Dipendra's face, must have carried out the killings. How else could the prince have killed so many people so quickly, he asked, reasoning that the actual prince must have been poisoned, and the gunmen, working for some conspiratorial power (Gyanendra's agents? the Maoists? Indian intelligence? the CIA?), had then eliminated everyone but those in Gyanendra's line.

When Dipendra died on the morning of June 4, having lain unconscious since the night of the shootings, Gyanendra became Nepal's third king within the space of three days. In response, demonstrators poured onto the streets, accusing Gyanendra and his son, Paras, of perpetrating a fratricidal massacre. Some tried to obstruct Gyanendra's coronation convoy on its way to Pashupatinath and the Bagmati River.

That day, while the country was still ensconced in a media blackout, the new king enacted a public security ordinance, granting home ministry officials the power to arrest or detain anyone on grounds of acting against national unity or national security. Despite the ordinance's shoot-on-sight curfew, protests continued. That evening, Gyanendra announced that he would establish an official commission to investigate the royal massacre in which the day's unrest had implicated him.[7]

By Wednesday, June 6, after three days of curfews and violence, an article by the leader of the CPN-M appeared in the Nepali-language newspaper *Kāntipur*. In it, he claimed that the royal deaths were the result of a conspiracy involving Indian and U.S. intelligence agencies.[8] A swift government response brought the immediate arrest of the newspaper's editor and two co-publishers.[9] Meanwhile, security checkpoints appeared all over the city, and police raided shops selling commemorative—but now potentially subversive—pictures of Birendra and Dipendra. Still in the United States and learning of events from a distance, I soon noticed the disappearance of alternative explanations for the massacre in e-mails and other correspondence from Nepal. "Let us know exactly when you're coming," a message from a close friend read in this period. "We have too much to talk about in person."

By the middle of June, the official massacre investigation produced its final, unsurprising report. Prince Dipendra, it said, was the lone killer of everyone who died in the palace on June 1. Adding little more to original foreign press accounts of the incident, save some details about the prince's drug use and choice of ammunition, the official report confirmed that the prince was probably reacting to an argument with his parents over his marriage wishes. An official set of facts was now in place, offered with appeals to national security and state continuity, and it was nested within a political atmosphere in which doubting the government, or dwelling on the details of what might have *actually* happened, was increasingly regarded as antidemocratic and dangerous. The escalating Maoist People's War loomed on the horizon as a growing national crisis that threatened to dwarf the incident at Narayanhiti.

Interwoven strands of the Nepali nation-state, the Shah kingship, and even aspirations for environmental change were at once challenged by the prince's alleged murderous rampage, and they were threatened by the specter of what might ensue if the government simply crumbled in the massacre's wake.[10] The People's War threatened to transform Nepal into a Maoist republic—a change that for many urbanites was equal to the state's complete undoing. Whether this inspired fear or hope was contingent on one's relative position, but it was largely dreaded by those with whom I lived and worked, who believed that they and their city had much to lose in a future Maoist state.

While the CPN-M had publicly demanded the elimination of the monarchy for years, political factions within and outside Nepal had grown frustrated with King Birendra's apparent hesitation to deploy the Nepali army against Maoist rebels. With Gyanendra now in place as king, the terms of engagement were likely to change, perhaps plunging the country into full-scale civil war.

Under these circumstances, to accept the official massacre explanation without question was to concede the truth to some larger, nationalistic interest in maintaining a narrative of unity, despite the reality of deep and violent divisions throughout the country. The very survival of the state, and its capacity to effect any sort of change in the future, were at stake. In a *New York Times* piece published a few months after the massacre, Padma Ratna Tuladhar, a Maoist government liaison, explained the stakes of sorting truth from fiction:

> We know now that we had a crown prince who was a drinker and a drug user, that we have a new king who was a smuggler, that his son is a killer, and that we have a government that is so corrupt that it is incapable of effective action. But we also know that the people of Nepal want their rulers to do everything possible to avoid a war. What they expect is that everybody, including the king, the government and the Maoists, put the massacre behind them and do whatever is necessary to bring Nepal to peace.[11]

Although official facts of the massacre ran contrary to what most of my informants and friends privately believed, those facts were also regarded as the sole acceptable truth if Nepal was to maintain the possibility of continued, albeit reformed, democratization, embedded in the centuries-old configuration of power relations that had long been the historical and symbolic anchor of Nepali national identity.

Dramatic, rapid political changes continued during the months that followed the massacre at Narayanhiti Palace. Familiar calls for reforms—including a new constitution, the resignation of Prime Minister Girija Koirala, and an end to corruption in government—were amplified, and overt resistance to the new king's legitimacy was gradually supplanted by deep anxieties about the Maoist insurgency. Within months, Koirala stepped down, and a new prime minister, Sher Bahadur Deuba, took office. To widespread public relief, Deuba announced that a cease-fire and the start of peace talks, would begin in August. The government then eased some Public Security Ordinance restrictions, allowing the Maoists to convene open meetings in the capital for the first time. In what seemed to be a sign of progress, the Maoists then revised their original list of forty demands. It changed to contain only three items: form an all-party government, hold a constituent assembly, and make Nepal a republican state. This appeared to be progress, in advance of a second round of peace talks, which convened on September 12, 2001.

Meanwhile, King Gyanendra remained hidden from the public eye and would not become highly visible until he appeared to fulfill the king's ritual responsibilities for the Dasain holiday in October.

A "NEW" NEPAL

Hope that peace talks would bring an end to People's War violence waxed and waned during the fall of 2001. By the end of October, Prachanda declared that the Maoists would no longer negotiate until an all-party interim government was formed, but he later issued a statement that seemed to suggest a new flexibility about the terms of the monarchy. The government responded by repealing the Public Security Ordinance and releasing sixty-eight detainees said to be Maoists.

Just as public anxiety over June's royal massacre was beginning to fade, news came of a helicopter crash in far western Nepal, killing the late Queen Aishwarya's sister, Princess Prekshya. This time, rather than tracing the public shudder through the Internet, I personally watched it reverberate through the lives of people I knew well in Nepal. I had returned to Kathmandu to continue my fieldwork on river ecology and had found the city in turmoil. A few weeks after my arrival, still settling into my host family's home in Satdobato, the phone in the hallway rang with call after call about the latest royal death. Host family members and friends sorted through news, rumors, and their fears, revisiting conspiracy theories several months old and, for many, now unequivocally confirmed.

My host family's home is located at the southern end of Kathmandu's Ring Road, where seven roads, or throughways, intersect. Built with Chinese aid assistance in the 1950s, the Ring Road marked, for a few decades, a social and spatial boundary between city and non-city space. When I first arrived in Kathmandu in 1991, the family's house, which sits just outside Ring Road along a pathway southwest of the Satdobato intersection, was surrounded by agricultural fields. At that time, the grey cement home and its gated compound seemed out of place, stark pillars of development in an otherwise green, yellow, and dusky beige patchwork of crops.

Jyoti Pradhan, her husband, and their two sons were the first in their extended household to leave the joint family residence and establish a home of their own for the nuclear family. Both members of middle-class Newar families, Jyoti oversaw the domestic sphere while her husband worked as an auditor for His Majesty's Government and later in private companies. The family was among the very first to construct a house outside the Ring Road in that area, along a lane that would later host new homesite after new homesite, and eventually be inhabited by Bahuns, Chettris, and Newars. While Newar identity, ritual practices, and family domestic life animated every aspect of life in the "old house," the extended family home in Sundhara, the Pradhans' new home, was in some ways a new cultural and spatial frontier—a separate and, in important ways, newly autonomous setting for building a modern home and family.

The benefits of independence as pioneers in a new suburb were not without a social cost, however. Tensions with neighbors inevitably reverted to explanations linked to their caste or ethnic origins; difficulties within the family would often arise over the stress of moving between the joint family household in Sundhara and the separate, nuclear home in Satdobato for religious and familial obligations.

For present purposes, it is important to notice that the family setting in which I lived was part of a new suburb, a social and built landscape that middle-class Kathmandu residents associated with modernity, change, and independence. But it was also a setting that, although once at the city's edge, was itself engulfed by suburban expansion during the 1990s and the early twenty-first century. Today, Satdobato is just one of dozens of dense residential neighborhoods outside of Ring Road in the south of the city. One must drive to find even one of the patches of agricultural fields that were the norm when I first arrived in the Pradhan home in 1991. From a place at the margins of suburbanization, Satdobato is now itself one of Kathmandu's older suburbs.

A few days after returning there in 2001, Jyoti sat with me to share early morning tea. Now in her late fifties, Jyoti identifies herself as a devout Hindu. She is the mother of two sons but has no daughters. Since I first appeared on her doorstep as a college student, Jyoti welcomed me into her domestic world and much of her sacred world, declaring me her "daughter." In over a decade of return visits, our relationship grew very close, and, indeed, deeply familial. When I returned in the fall of 2001, I found a city in the throes of uncertainty and a home filled with unusually deep anxiety.

Over our morning tea, Jyoti's voice hushed as her tone turned mournful and serious. "How could the prince have killed all those people?" she asked me. "There's no way he could have done it alone. We may never know who it was, but we do know that they now have power; they are the ones really ruling the country." Just as she was saying this, the phone rang. Sarita, a family friend, was phoning to tell us about Princess Prekshya's death in the crash. "It's not over yet," Jyoti said, clutching my arm as I hung up the phone. "More people will die." In that moment I couldn't discern whether Jyoti's anxiety was born of foresight or panic, and in the weeks ahead I puzzled over the meaning that friends and informants assigned to the massacre and its enduring primacy in their national consciousness.

For months after my arrival, I found myself engaged in repeated conversations about the transformation of Nepal's kingship. It was clear that the monarch's symbolic moral authority was also wounded in the massacre, and this seemed to produce a combination of fear and defiance. "We show respect to the king in practice," Jyoti told me of King Gyanendra, "but we no longer follow him in our heart and mind." Only my informants who relied exclusively on the foreign press, such as expatriates in the development community, and my family friend, Akash, whose nascent career in the Royal Nepal Army required unquestioning loyalty to the king, seemed to believe the official story of how Birendra had been killed. For nearly everyone else in my network of friends and research associates, respecting the new king seemed utterly impossible.

The official version of massacre events was repeatedly described to me as simply out of the question—a lie that the government might be able to tell the rest of the world but not to the Nepali people, who knew the nation-state's history. The question was not whether Birendra and his family had been murdered in a plot more complex than in the official story, but *how* this could have happened, and how to reconcile the accounts of witnesses with the plot as it was imagined. When witness testimony still failed to square with alternative explanations, my informants sometimes dismissed witness accounts altogether as lies told under duress. They reasoned that witnesses were controlled by the same forces that killed Birendra's line, so naturally they spoke out of fear for their lives.

Rakesh, a middle-class Bahun family friend, explained a rumor in which a plastic surgeon made Dipendra appear to have shot himself. He called King Gyanendra *chalāk* (clever, cunning, not necessarily trustworthy) and told me that, privately, Rakesh's friends were saying *marhos* (let him die). Even Sama, who just one year before at Pashupatinath, had confided her trust in the king's power and moral authority, believed it was Gyanendra who had directly orchestrated the massacre. She speculated that several perpetrators simultaneously committed the shootings, but each had worn a mask in order to look like the prince. This way, she reasoned, eyewitnesses only saw who they thought was Prince Dipendra, while in fact there were several killers. She also believed that the chief of the army was involved in the conspiracy.

Others told me that people living in the neighborhoods surrounding Narayanhiti Palace reported hearing helicopters coming and going on the night of the massacre, while some argued that palace guards were sufficiently armed to have at least controlled the incident and prevented such widespread bloodshed. Still others compared the weapons reportedly used by the prince and the time that elapsed during the incident, arguing that the two could not be reconciled. Such speculation pointed either to official details that seemed contradictory or to blatant omissions that rendered the official story perilously incomplete.

As a result, Jyoti told me, people didn't honestly honor or follow [*māndainan*] King Gyanendra. Any lack of public discourse or dissent surrounding the facts of the royal power shift was due to fear—it had nothing to do with acceptance of the new king as being legitimate. This was hardly the ultimate moral authority that Sama once hoped would make river restoration possible. Fundamental transformations in the urban middle class regarding the kingship, the polity, and the cultural idea of the state were underway, and the form they would eventually take would undoubtedly affect the terms of future change, environmental or otherwise. Whether that form would be a differently functional democracy, a Maoist republic, an authoritarian state, or something else, and whether any of these would retain a place for monarchy were active and open questions.

LOSING MONARCHY

But what would it mean for these urbanites, and the wider Kathmandu populace, to lose their king? How did monarchical power come to be such a central feature of state symbolic and material power? How would a changing monarchy alter the state apparatus and its capacity to make change? How would it impact the symbolic place of the Bagmati riverscape in the stories of the city's past and aspirations for its possible future?

The emergence of kings and eventual organization of the general territory that is now Nepal around a Hindu monarch can be traced to political unions formed by villagers in an effort to protect their lands (Stiller 1999; Burghart 1996). These unions required authoritative figures to settle disputes, and the local leadership that resulted is believed to have led to the kingship. The kings of Nepal's Shah Dynasty were descendants of the king from Gorkha, a principality in central Nepal that interacted largely with what was at that time the League of Twenty-Four Kingdoms in the Central Himalaya.

Vedic texts and Puranic royal rituals call the king a "rainmaker, guarantor of the cosmic order and bride of the earth" (Bhatt 2002:223; Hoeck 1990; Whelpton 1992:9), and popular accounts in Nepal long held the Shah king to be a reincarnation of the god Vishnu. The precise origins of this idea are contested; while Whelpton draws on the work of Toffin to trace this belief to the fourteenth century, others note a seventeenth-century *vāmshavalī* description of the Gorkha King Rama Shah as *vishnuko amsh* (a portion of Vishnu) (Bhatt 2002:223). Successive Shah kings maintained and expanded their power by cultivating patronage networks through vast land grants, and they formed a state bureaucratic apparatus that favored high-caste groups and the Newars of the Kathmandu Valley. Throughout the kingdom, they encouraged ideas of divine kingship and monarchical moral authority, and concretized their symbolic powers by retaining ultimate legal and military authority (e.g., Burghart 1996; Dirks 1990; Dumont 1970:62–88; Heesterman 1985; Inden 1978; Whelpton 1992).

Regardless of its origin, popular regard for the Shah kings as being the possessors of extraordinary power persisted in Nepal through the twentieth century. According to Nina Bhatt, a scholar, this was evident in the aftermath of the royal massacre, when "Kathmandu youth mourning the loss of Birendra told [her] that while they themselves were somewhat skeptical of the idea of the king as Vishnu, they did recall that their fathers would offer food to the king alongside other household deities before partaking of their own meals" (2002:224). This everyday ritual occurred in the Pradhan household, as well as some of the homes of my research associates and friends.

My host family took part in another mundane reproduction and affirmation of the monarch's important status by prominently displaying *rajaraniko photo* (pictures of the king and queen) in public and more private areas of the household. Jyoti explained to me the purifying power of beholding the face of the king, ritually restated annually when, on the king's birthday, one could stand in line at the palace to do precisely that (Bhatt 2002:224; Whelpton 1992).

Recognition of a divinity-like power and moral authority was the complex product of years of nation building and official efforts to cultivate affective ties to the Nepali nation-state. Constructed as a protector and caretaker in history books (Onta 1994; 1996), in religious practices (Burghart 1996), and through an official history that placed Shah kings at the center of heroic accounts of the country's unification, the king was to have moral purity and national heroism, which were important features of the narratives through which Nepalis were encouraged to think of themselves as a single, united nation. Pride in Prithvi Narayan Shah in particular, rests for some on the idea that he was both a unifier and a protector from Mughal and British colonial rule. Thus, the power of his idea of an *asal* Hindustan—a "pure" country of unity despite tremendous ethnic, caste, linguistic, and geographic diversity within its own borders. More comprehensive approaches to this conventional nationalist discourse question the extent to which Prithvi Narayan Shah was actually the heroic multiculturalist that he was purported to be within Nepal, suggesting that Prithvi Narayan was more concerned with "[preventing] his kingdom from becoming a garden of every sort of people, [as] only then would it remain a true [*asal*] Hindustan of the four *varnas* and thirty-six *jats*" (Bhatt 2002:226; Gellner, Pfaff- Czarnecka, and Whelpton 1997:24).

Regardless, successive Nepali kings presented themselves as promoters and protectors of Hindu dharma, particularly in relation to

Mughal rule in India (Burghart 1996). While fraught with contradictions, as many scholars have shown (e.g., Bhatt 2002:225; Regmi 1975:221–24; Whelpton 1992:13), the social and cultural product was a polity in which the power to make change was formally, and sometimes popularly, bound up with the monarch.

THE KINGSHIP AND DEMOCRATIC MODERNITY

After 1951, the kingship changed in articulation with the distinctly Nepali experience of shifting between absolute monarchy and multiparty democracy. After the Ranas, who had ruled since 1846, were overthrown, King Mahendra introduced his experiment with multiparty politics. This was short-lived, however, and was abandoned in favor of governance systems ostensibly more "suited" to Nepal. Burghart explains,

> During the next decade Nepal embarked on an "experiment" in parliamentary democracy, which King Mahendra decreed in 1960 to have been a failure, and which he replaced with a new system of government that was composed of the king and four tiers of elected councils called *panchayats*, which were constituted at the village, district, zonal, and state levels of government. According to Mahendra, the panchayat system and the kingship were the traditional forms of government in Nepal; these two political institutions were especially suited to promote unity and development in the Nepalese context. All sovereignty, however, was vested in the kingship, and the panchayats were granted only an advisory function in the government of the kingdom. (1996:256)

A new constitution instituted the panchayat system, whose ruling apparatus was partly elected and partly appointed. Political parties were officially banned because they were contrary to Nepali tradition and culture, while panchayats were officially promoted as being more democratic than the elite democracy practiced in the 1950s (Burghart 1994, 1996; Hoftun 1993; Hoftun, Raeper, and Whelpton 1999:xii). Most importantly, the palace held central control over all processes of governance. This system remained in place, with minor alteration, until the jana andolan led to the reinstatement of multiparty democracy. A dramatic decentralization of power would assume its full form only in the decades ahead.

In the aftermath of the royal massacre, it was impossible to fully separate sentiments of mourning the royal family from disappointed assessments of the post-andolan period of democratization. In fact, Bhatt argues that "the link between the deaths of the royals and the dysfunctional state of governance was the most striking aspect of post-massacre discourse. . . . Ordinary Nepalis viewed the massacre, specifically the assassination of the King, as a sign that Nepalis were not intended for democracy. . . . The popular reasoning was that if the monarch were still in power (as in pre-1990) then the massacre could never have happened." She offers ethnographic elaboration of this point by recounting the reactions of her informants; among them, a middle-aged Chhetri man told her, "In a country like ours, only the king's rule works [*hamro deshma raja ko shashan matra chalcha*]. No one is educated. Those in Parliament eat our development budget" (2002:231–33).

The level of distress, uncertainty, and fear that Bhatt's work captures resonated in my host household and among those with whom I worked. It would only intensify after the massacre, as an already disappointing democracy became a national state of emergency and as expectations that the war between the government and Maoist rebels in the countryside would reach the capital city mounted.

AFTER THE MASSACRE, AN EMERGENCY

Five months after the massacre, a third round of peace talks between the government and the Maoists broke down, and Prachanda announced that the cease-fire was off. A few days later, the Maoists resumed nationwide attacks, with offensives in Surkhet, Dang, Syangja, and Salleri. Government casualties included—for the first time in the history of the conflict—at least fourteen Royal Nepal Army (RNA) soldiers. Fifty police officers and several other officials were also reported killed. The number of Maoist casualties was unclear; although as many as sixty were reported, only fifteen bodies were recovered.

In response, Prime Minister Sher Bahadur Deuba announced a state of emergency on November 26, 2001. The government mobilized the RNA to fight the insurgents, who Deuba officially declared as "terrorists." The cabinet enacted a Terrorist and Disruptive Activities Ordinance, authorizing arrests without due process and facilitating new controls over media and information.

Although we were aware of its inevitability, my host household learned of the emergency through an announcement in the Nepali press, here in its printed English translation:[12]

> His Majesty King Gyanendra Bir Bikram Shah Dev, on the recommendation of Council of Ministers, declared a state of emergency throughout the country effective from Monday, the Royal Palace announcement said. The King has authorized the deployment of the army on the recommendation of the Council of Ministers, a Defense Ministry spokesman said. His Majesty declared the state of emergency exercising the authority given by Article 115 of the Constitution of the Kingdom of Nepal. His Majesty the King has suspended sub-clauses (a), (b), and (d) of Clause (2) of Article 12, Clause (1) of Article 13 and Articles 15, 16, 17, 22 and 23 of the Constitution, except the right to file habeas corpus. The Council of Ministers in its emergency meeting decided to recommend to His Majesty the King to declare the state of emergency throughout the Kingdom to control Maoist attack and ensure law and order in the country.
>
> His Majesty the King, on the recommendation of the Council of Ministers, also issued the Terrorist and Destructive Activities (Control and Punishment) ordinance in order to contain terrorism and violence in the country. Meanwhile, the government has declared Nepal Communist Party (Maoist), its sister organizations and any organization and individuals that support Maoist party and its activities as terrorists. The decision of the government came following recent violent attacks by Maoist rebels in different parts of the country in which more than 70 policemen and army personnel were killed in just three days.

Like news of other recent events, my host family received this announcement with a combination of fear and suspicion. For much of the day, Jyoti talked on the telephone, nervously consulting relatives and those she knew with connections to the military or civil service.

As with the massacre, I wondered what the emergency meant to my informants. How would they interpret this event and assign it narrative coherence? The very processes through which social, political, and environmental change might take place would be shaped, in part, by the meanings and consequences of this pivotal political development.

The controls enacted by this emergency were only new in the context of Nepal's recent past. Emergencies followed the suspension of democracy in 1959 and the jana andolan demonstrations in 1990. During the Panchayat period, rights to assemble, publish, and freely express political opposition were nonexistent, so the political reality in which my older informants spent their lives before 1990 was one characterized more by secrecy, evasive speech, and the everyday risk of being branded an antinationalist, than it was by free expression or open debate. Furthermore, despite years of nominal free press under post-andolan democracy, it was difficult to assess the extent to which public faith in the press as being independent or impartial had fully developed, and among what kinds of audiences.[13] The emergency was thus in some ways a reversion back to a distant but familiar time when "democracy" simply signaled a certain form of Panchayat governance, characterized by a tightly restricted political climate.

Yet references to the past and its precedents were noticeably absent. Rather than invoke history to make sense of the emergency, my host family and friends insisted that nothing quite like what we were witnessing had ever happened before. At stake, I was told, was the survival of the very nation-state itself. Monarchical upheaval, intensifying violence in the countryside, and the restrictions imposed by the emergency signified a wholly new political moment, a completely "new" Nepal.[14] The "old Nepal" became post-1990 *prajatantra* under King Birendra, used, for instance, one morning when, unprovoked, Jyoti said, "Look at how different the old Nepal [*pahileko* Nepal] and the new Nepal are. At least then we had a king. They say it was his son, but they're lying to us. Everything since then has been lies." In this and other comments, Jyoti offered the emergency as evidence that the monarchy since the massacre exercised little, if any, effective symbolic authority. It was precisely the king's legitimacy that separated an old and new Nepal. Not only did the emergency foreground military rather than symbolic means for claiming that legitimacy, then. It also consolidated for this member of my middle-class network of family and friends the fundamental popular doubts and fears raised by the massacre.

In the days following the emergency declaration, stories of deadly military clashes filled the news. Thirty-four people were reported dead in Solukhumbu on the first day; the next day, sixteen. Socket bombs and pressure-cooker bombs started appearing in bazaars inside the Valley; a bomb exploded in a Kathmandu carpet factory. Throughout the emergency, which would last for ten months and for which there were 4,300 officially reported war deaths, it was difficult to discern what was actually happening through the haze of controlled information, rumor, and raw fear. Yet again, to do so — to assign a narrative to the confusion, denial, and anxiety — was to maintain a sense of belonging to the capital city and to the larger nation-state. In a new Nepal, Kathmandu was a newly fragile refuge — for the state, its privileged subjects, and those fleeing for their lives from the countryside.

PRIVATE DOUBT, PUBLIC TRUST, AND "OUTSIDE" STORIES

The 2001 emergency involved explicit government censorship that was justified on civil security grounds, and this was argued to be a critical official step in the effort to deprive the Maoist Movement of a legitimate public platform.[15]

At home in Satdobato, we relied on a combination of FM radio and state-run television news for daily information. There was no computer in the house, and hence no Internet access without a journey to a commercial Internet shop in the Patan tourist district. Although I did this fairly often, it was not the family's primary mode for gathering daily news.

Jyoti and Krishna regularly tuned into an early-morning news-interest FM radio program called *Hālkhabar*. In the evenings before dinner we watched the state-run television news broadcast together, huddled under blankets or in front of a small space heater. These everyday habits of news gathering continued through the emergency, despite our knowledge that media-control provisions censored everything we saw and heard. To tune in, of course, was not the same as to accept the reports as being true, and family members regularly and explicitly distinguished between consuming official news and believing it.

The tension between information and its official confirmation in state media was made particularly poignant a few days after the emergency announcement. On November 29, 2001, Sama called to ask if I had heard about a bomb blast that morning at the Coca-Cola plant in Balaju Industrial Estate on the northern end of Kathmandu. The radio news had not reported it, she told me, but her mother heard the blast from the family home in Baneswor, a considerable distance from Balaju, and Sama's husband had learned of the incident a few hours before at work.

Sama said that she did not expect this to appear in the news immediately—if at all—explaining that, because it was the possible work of Maoists inside the Valley, it would signify yet another government weakness. Jyoti watched me hang up the phone with a worried look. "Please don't go out today," she said. In the unknown of a city in a state of emergency, it was impossible to assess whether Jyoti's advice should be interpreted as paranoia or the voice of cautious reason, so I rescheduled a set of meetings and agreed to stay home.

When I phoned a colleague to cancel our plans to meet, I explained that Jyoti had asked me to avoid going out because of the bomb blast earlier that day. I mentioned that I thought the silence about the incident in the news, perhaps even more so than the incident itself, had caused concern and compelled Jyoti to encourage me to cancel my plans. My colleague, an expatriate development professional, was confused, suggesting that the concern was at least an overreaction, at most absurd. "But it *is* in the news," he said. "It's been on Reuters and CNN all day. Actually, no one is even sure if it was the Maoists."

I hung up the phone, tuned the Pradhan family's television to CNN and saw the news of a bomb blast in Kathmandu scrolling across the base of the screen. I called to Jyoti and showed her the headline, but to my surprise she was not relieved. Until it was confirmed in the Nepali news, she explained, it could be dismissed as *hallā* (rumor) or *bāhirako* (outside) stories. She then volunteered that this was just like waiting for news confirmations during the royal massacre that past June, when official sources were silent for hours even as everyone with access to satellite television was already aware of events at Narayanhiti.

Because of the way international news sources figured into the royal massacre story, Jyoti's reaction to a globally circulating report seemed counterintuitive. My inclination was to trust the international press,

110 Chapter Three

since we all knew that the emergency ensured an overtly censored national newscast. But watching for state confirmation was, in the end, not motivated by the goal of confirming the facts themselves, but rather to learn which facts were officially acknowledged. Until the state news reported the bomb blast, it was entirely possible that a very serious incident was again being deliberately withheld from the public, a condition to fear because there may, then, be others. On one level, Jyoti was gauging the emergency itself: She wasn't questioning the blast, but she seemed to fear knowing about it before it was "officially" true. Official refusal to cover the story might signal that the government itself was legitimately scared, in denial, or acting unrealistically.

This points to a tension between circulating stories, the confirmation of those stories in state-controlled news, and an overall skepticism about all state news narratives. A Maoist bombing at the Coca-Cola plant was finally confirmed, but a few days later Sama told me that she heard that it was in fact not the work of Maoists after all. In the end, it was never clear who or what had set off that explosion or many others that followed in the capital.

After this episode, Jyoti spoke with me about the tensions between truth, doubt, and aspiration that punctuated every effort to figure out what, exactly, was going on. "These days, we say things that we don't actually believe in our hearts and minds [*ājkāl mukhle bolcha tara manle māndaina*]," she told me, marking an explicit discordant relationship between private knowledge, doubt, and fear and public performances of loyalty, trust, and hope. What was known and believed could not always be said, and what was said could not always be openly known or believed.

CLEAR FLOWS THE BAGMATI

Through the blur of doubted news reports in this period came a differently unbelievable account of miraculous change, best captured in the striking headline "Clear Flows the Bagmati."[16] For the first time in recent memory, the reach of the Bagmati River that flows past Pashupatinath Temple was suddenly clean and clear, the result of a 522-meter-long, 2-meter-wide diversion tunnel. Nearly ten years in the making,[17] the tunnel channeled all river inputs above Gokarna (mainly from Mitra Park and Gokarna) to a drain below Tilganga. The effect was to "clean" the Bagmati reach that flows past the temple and its cremation ghats. For a few months, the diversion tunnel channeled raw sewage inputs from one end of the complex to another, but it was later connected to a sewage treatment plant at Kumarigal, near the Guhyeswori Temple. The effluent would now be discharged back to the river only after passing through a water purification system.[18] Not only was this an apparent development success, but its site was the Bagmati, the degradation of which presented a most intractable bundle of problems on which billions of rupees had been previously spent without results.

Five months after the massacre, and just a few weeks after the diversion tunnel was opened, I traveled again with Sama to Pashupatinath. We walked the riverbank quietly, watching and listening to the bustle of puja activity, and, with many others, we marveled at the seemingly clear river water. Standing across from Arya Ghat, Sama broke our somewhat stunned silence by recounting, in detail, her recollection of the royal cremations. She told me that she could still feel the shock of the days after the massacre, and she still had vivid mental images of the royal family's last rites. Rather than marking a new beginning, the most recent story of royal succession had brought about unprecedented turmoil. And yet, in a strangely unforeseen way, in its aftermath there actually was a change in the Bagmati's condition—however small, insignificant, and perhaps temporary. We had come, after all, to see the newly cleaned river reach—to see and believe it for ourselves.

The historical, cultural, and mythical importance of Pashupatinath infused this minor development project with powerful symbolic meaning. Its timing was so curious, in fact, that it was almost impossible to reconcile the fact that it was ten years in the making. Surely the site and timing were demonstrations of strength by King Gyanendra and the state newly under his charge or of the postsuccession transformation Sama had predicted, albeit delayed.

Perhaps most striking was the narrative of how the "miracle" was achieved. The project was administered through His Majesty's Government and the Pashupati Area Development Trust, rather than the usual "cooperative" effort between international or bilateral development agencies and HMG. Its relative autonomy gave the appearance of an independently strong and capable development apparatus free of donor influence, in sharp contrast with its pre-emergency opposite.

Following that day at Pashupatinath, Sama and I returned for several visits to the offices of the Bagmati Area Sewage Construction and Rehabilitation Project (BASCRP), the government agency responsible for the planning, construction, and operation of the diversion tunnel and treatment plant.[19] We met with the project managers, who recounted the project story with great pride.[20] One emphasized that this was the first project of its kind to be designed and financed "internally," without foreign "experts" or foreign investments. He stressed that the project provided practical evidence that the Nepali government could accomplish "development" free from the constraints and dependencies associated with international donors. This made the BASCRP remarkable as an apparent step forward in the quest to improve water quality in the Bagmati and Bishnumati system. It demonstrated Nepal's "independent" capacity to undertake environmental management as development. We were repeatedly told that the state was committed to cultural preservation and the natural health of the Bagmati in the capital city. Nationalist pride in state autonomy, and the deeply historical notions of purity that they invoked, formed the subtext of the "miracle" at Pashupatinath; its location and timing only strengthened its symbolic power.

But purity and autonomous power would need assistance if they were to be scaled to fit the entire river system. A senior project manager explained that BASCRP success brought with it expectations of replication. This single plant could not stand alone, and building treatment facilities at other sites, admittedly critical to the health of the Bagmati as a system rather than as a series of isolated reaches, *would* require outside development assistance. A full-scale river rehabilitation effort required a network of treatment plants, which was simply too expensive for HMG to construct alone. The ultimate aspiration was to complete the ecological rehabilitation of the entire Bagmati,[21] the manager told us, but this could only be done with international assistance.

He then emphasized the importance of Nepali autonomy, even as he defined that autonomy's limits. He called the riverbank sukumbasi presence a major obstacle to building additional sewage treatment facilities that made the challenge of securing international funding much more complex. Donors, the manager noted, would likely require that squatters be resettled, and the government was ill prepared to do this on a scale equal to what was needed.

Over several visits, we were guided through the sewage treatment facility at Guhyeswori, learning its technical features and hearing stories about plant construction. While happy to emphasize the BASCRP's success, the managers with whom I spoke also recognized the limits of this localized remedy in a broader mosaic of systemwide problems for the river. In addition to broader questions of the river system and restoration funding, there was the question of individual facility maintenance. Industrial chemicals were regularly released into the river upstream, an ongoing complication in the water treatment process. There was already evidence that certain critical bacteria levels were unstable as a result.

Indeed, within a few months, accolades for the miracle at Pashupatinath were replaced in the press with stories of its looming failure. On February 8, 2002, in the Nepali-language *Kāntipur*, an article listed at least sixty-eight industries whose inputs upstream of the Guhyeswori plant actively threatened to destroy its water treatment capacity.[22]

The diversion tunnel remained in place, ensuring relatively clean and reasonably plentiful water in the Bagmati at Pashupatinath. Almost unimaginably, and yet in strange accordance with Sama's prediction, we gazed at a "clean" Bagmati from the Arya Ghats, but its specificity made it more a performance of change than a lasting and impactful enactment of it. Even as its Pashupati reach flowed clear, then, river degradation continued. Leaving the temple complex that day, Sama told me that, despite having witnessed her prediction fulfilled, she had lost her hope for real and lasting river transformation.

URBAN RIVERS IN A NEW NEPAL

Long before the turmoil unleashed by the royal massacre, confronting river degradation evoked passionate expressions of loss and despair. If, for Sama, the monarchy once possessed the ultimate capacity to restore the Bagmati, others, despite a mounting tally of unmet expectations, anchored their hopes to aspirations for democracy. The royal massacre was felt among those with whom I lived and worked as nothing short of a breaking point, explicitly captured in declarations of a new Nepal. In the midst of deep panic, confusion, and frustration over blurred boundaries between truth and fiction, informants described their private doubts, suspicions, and fears, and they contrasted these with public, and necessary, performances of loyalty, belief, and trust. There seemed to be no question that the new king was illegitimate and had, in some ultimately unexplainable way, managed the massacre of most of his family. As a result, his immediate moral authority was unquestionably lost, and its future was in serious doubt.

But another threat loomed larger than a king's lost legitimacy. The CPN-M, representing the countryside, the marginalized, and the potential for complete transformation of the nation-state, instilled panic among my informants that surpassed even that evoked by the new king. Those who feared a Maoist state, and the bloody struggle through which they imagined it would come to pass, claimed to have no choice but to align with a government they mistrusted. The stakes were clear: The sociopolitical order that had for centuries privileged the Valley, the city, and its elites, was for the first time in the history of modern Nepal formidably and seriously threatened.

The emergency deepened urban anxiety over the survival of the state apparatus itself. With absolutely no assurance of reasonably factual news coverage, and with the paltry gains of a decade of democracy essentially revoked, Kathmandu residents were effectively captive in their Valley refuge, temporarily safe from the overt violence of the war on one hand but confined to a historically familiar reality of controlled information on the other.

Amid censored news, rumors, and deep suspicion of the new king, official pronouncements of a clean Bagmati River seemed at first utterly absurd. Yet we found precisely that—an apparently clean, clear reach of river flowing just past the Arya Ghats. Its symbolically charged location and powerful story of autonomous, donor-free development gave it a logical place in a longer history of the nation-building narratives discussed in chapter 2, but its partial and temporary character dulled its potency. Here, in this place where royal power and the nation-state itself were ritually reinforced over time, "restoration" in a new Nepal had begun.

What, precisely, would urban ecology look like in this new Nepal, in which a contested and distrusted kingship, a revolutionary army, and a state military apparatus framed the political conditions of the present, and shaped, through overt violence, the possibilities for change? How would the meaning of urban ecology in Kathmandu change when the kingship was no longer regarded as an ultimate arbiter of moral order, or when the democratic apparatus through which change was framed for the past decade was eliminated?

If we are to understand ecology as a future-making strategy, a way of setting and experiencing conditions for urban and social transformation, then let us turn to the river-concerned actors who anchored their hopes, desires, and actions toward ecological change in this volatile period. What would emergency urban ecology prioritize? What would urban nature become in this moment, and what would be deemed the right and proper way to relate to it?

4
Emergency Ecology and the Order of Renewal

ALMOST IMMEDIATELY AFTER the emergency was declared on November 26, 2001, citizens of the capital city witnessed a massive urban beautification campaign. Housing demolitions took place in several locations, and new urban parks were built with astonishing efficiency. Despite the coincidence of their occurrence with the emergency, these environmental changes were officially described as preparations for an important meeting of regional dignitaries, scheduled to take place in Kathmandu in January 2002. This, the eleventh meeting of the South Asia Association for Regional Cooperation (SAARC), convened at a time of escalated tensions between India and Pakistan, and as such it attracted unusual attention from international media.[1] Nepali government authorities framed the Kathmandu beautification measures as an integral part of ensuring an appropriate welcome for SAARC delegates.

Yet, in the context of an emergency, which many urbanites regarded as strong evidence that the violence in the countryside had become a critical threat, the demolitions and rapid construction of parks also demonstrated official order and control. For a state now fighting the Maoists for its very survival, the demolitions and parks were symbolic assertions of national territorial rule. Kathmandu was, in fact, still the center of the nation-state of Nepal, however contingent that state had become in the turmoil of the royal massacre and the People's War. A near perfect overlap of the emergency and SAARC beautifications facilitated, on some levels, a conflation of the two. This chapter explores the dynamics of urban environmental change during the emergency, and it highlights how environmental space was engaged and shaped toward specific political ends.

On one level, emergency beautifications were unsurprising for a state seeking to retain its local and regional legitimacy. Such environmental transformations are not unusual strategies for governments seeking to avert crisis or bring order to uncontrolled areas (e.g., Greenough 2003; Sivaramakrishnan 1999; Tarlo 2002). While the Maoist army may have remained beyond the government's reach, neglected and largely informal city spaces were not.

But an appreciation of the relationship between urban environmental space and political aspiration in this case requires a look beyond the emergency as well. Sites of demolition and park making did not remain the exclusive domain of authoritarian rulers, and their message of state strength did not endure. Nearly five years after the initial beautifications that I describe in this chapter, with the monarch's control significantly weakened, citizens across an unprecedented social spectrum united to demand the deposition of the king. The same parks that appeared, and were welcomed, as evidence of state effectiveness and resilience in 2001 became sites from which to amplify state failure and to demand reform by 2006.

The massive civil unrest and antigovernment, indeed antimonarchy, protests that followed the emergency assumed a size and scale unseen in Nepal's history. Eventually, they brought about the reinstatement of democracy and the near-total elimination of monarchical power and privilege. Urban environmental spaces constructed during the 2001 emergency were transformed again in this next era, but at that time they became staging grounds for performing new visions of a nation and polity *without* monarchy.

This chapter begins with, and focuses primarily on, environmental initiatives undertaken by King Gyanendra's government in the weeks after the emergency declaration. But to highlight their dynamism, I conclude with a discussion of one park, Maitighar, which was transformed from a symbolic restatement of monarchical sovereignty in 2001 into a locus of citizens' claims to a new postmonarchy Nepal in 2006.[2]

REASSERTING THE STATE
THROUGH URBAN ECOLOGY

In late 2001 and early 2002, the complex combination of violence related to the People's War, the upheaval of the monarchy, and the declaration of the emergency seemed to consolidate a growing sense of uncertainty, contingency, and explicit fear in public discourse in Kathmandu. The legitimacy of King Gyanendra was held in active question in the wake of the royal massacre, and the government was increasingly perceived to be vulnerable to defeat by the Maoists. These conditions held unknown consequences for the urbanites with whom I worked, who often expressed a fear of "losing Nepal" to the Maoists.

Fraught with public anxieties, the capital was about to host the SAARC summit, which was itself framed by extreme regional and global tension. Relations between India and Pakistan had approached a dangerous breaking point after an attack on the Indian Parliament on December 13, 2001, in which twelve people were killed and twenty-two were injured. Officials in the Indian government immediately blamed Pakistani militants. Internationally, still-fresh pronouncements of the "War on Terrorism" by the United States gave new global currency to the category "terrorist," reshaping political dynamics throughout South Asia.

From December 2001 through January 2002 in Kathmandu, I found it nearly impossible to differentiate between SAARC preparations and the general tightened security that characterized urban life under the emergency. Early one morning, I arrived in the core of the city, at the bustling intersection at Thapathali, to find each of the colorful billboards that crowd this area draped in enormous white sheets. Normally, the boards advertised products like Wai Wai Instant Noodles, Royal Stag Whiskey, and Shikhar Cigarettes; now they stood stark and blank, rising ominously above the winter fog. I puzzled over an interpretation of this whitewashing at first, wondering if it could be a new form of control over advertising related to the emergency. I soon realized that it was more likely a feature of SAARC preparations. Indeed, over the next few days, Kathmanduites watched from the streets below as painters adorned the fresh canvases with SAARC-related welcome slogans, painted in uniform, all-capital English lettering. Pronouncements like "LONG LIFE TO THE SAARC PARTNERSHIP!" hovered all over the city.[3] One could scarcely travel a main thoroughfare without encountering such neatly refashioned "ads." The air of festivity and welcome created by the SAARC signs blended almost eerily with their ubiquity. Slogans reminded urbanites of the upcoming SAARC conference, but the billboards on which they were painted were ultimately the domain of the state that would host the conference. That state had just assumed a new kind of authoritarian control, under which no change would be enacted on a "democratic" or consultative basis.

In the first few weeks of the emergency, the streets of Kathmandu buzzed with environmental improvement projects. Trees were planted, thoroughfares were greened and painted, and new parks appeared, seemingly overnight. The phrase "SAARC *banaune*," or "making SAARC," became a common explanation among my friends and research collaborators for the massive, rapid urban transformation. "Park *banaune*" soon joined the phrase, and the idea that "SAARC *banaune bhaneko* park *banaune*"—that preparing for SAARC is defined by constructing parks—functioned at once as a statement of fact and, often, as an expression of astonishment at the pace and entirety of Kathmandu's environmental transformation. This phrase conveyed how the coincidence of beautifications and SAARC provided logic to authoritarian environmental interventions, activated a particular kind of consent, and enabled a form of governance quite at odds with the city's and nation-state's recent history of democratization.

Dramatic changes took place all over the city, often with a rapidity that seemed akin to spontaneity. Over little more than a week, traffic islands in Patan and Kathmandu were freshly planted. Where cement road dividers did not previously exist, they appeared almost instantly, constructed quickly by huge groups of day laborers. Long lines of workers painted sidewalk edges in black and white stripes, while tall welcome gates, decorated with flowers and graced at the base with *karuwa*s (rounded brass pitchers), were erected at key points on major roadways.

In the simultaneous imposition of the emergency and preparations for the conference, I sought to understand seemingly contradictory performances of festivity, openness, and "a hearty welcome" on one hand, and forms of local control and silenced dissent on the other. The physical transformation unfolded against a backdrop of militarized streets, lined with young army personnel who wielded rifles or machine guns and were often clad in riot gear. Traveling around the parliamentary offices at Singha Durbar entailed passing through gauntlets of soldiers with guns poised, fingers resting on triggers. It was clear that the welcome signified by SAARC preparations was intended for specific regional guests only; those who would inhabit the city long after dignitaries left were well advised to notice the soldiers in equal measure. And yet the combination of rapid park making and heavy military coverage gave a strangely reassuring impression that everything, from environment to society, was under the eye of a state that would protect, safeguard, and even "green" its urban capital territory.

ENVIRONMENTAL CHANGE IN THREE SITES

A variety of measures characterized official SAARC preparations in Kathmandu, but three most visibly intersected with my broader research project on river improvement. First, there were swift building demolitions and new park construction at Tinkune, an area of land in the vicinity of the national airport. Second, the four- and five-story buildings at a crossroads called Maitighar were demolished and, over the course of a few days, were replaced by a small park featuring an enormous mandala, a small *stupa*, and two stone water spouts.[4] Finally, just below the Bagmati Bridge, the residents of two informal settlements that were officially considered illegal were forcibly evicted and the homes were leveled. There was a notable lack of public protest in all three cases.[5]

Control over the development of Tinkune, a stretch of land not far from the city's airport, had been in dispute at least since 1974, when the area was first officially designated for urban parkland. A park was never built, however, and in the mid-1990s, the Supreme Court of Nepal ruled that the government had not adequately completed land acquisition procedures. The approximately fifty-five residents of Tinkune were convinced that their long dispute over the territory was settled and that they had assumed legal landowning status. Although fully built up, the lack of an official urban plan gave Tinkune the tenuous status of a semiformal settlement.

The emergency suspended the ownership rights that the courts had affirmed, and this, combined with an official desire to divert media attention away from the Maoist insurgency in the weeks leading up to SAARC,[6] precipitated the swift mobilization of park construction at Tinkune in December 2001. Residents and businesses were given fifteen days' notice to vacate the property, and they were offered compensation that was well below market value. There was little public resistance, however; a newspaper article explained the lack of protest by reasoning that, "in 'normal' times the Tinkune episode would have attracted much political opposition. But the pressure of pre-SAARC beautification and the Emergency means these are not normal times."[7]

Over the course of a week, this highly visible site was then transformed from an unplanned residential and industrial area to a lushly planted park with a large, dramatic pond at its center. When dignitaries from all over South Asia arrived for the SAARC conference, they drove by the new park on their way from the airport to the city center. In contrast to the history of dispute and alternative use plans that characterized Tinkune, the prior planning history of Maitighar, a much smaller piece of land, was unclear. Many press accounts claimed that there was no prior official plan for a park there, but the mayor made claims to the contrary in an anonymous *Spotlight* piece called "Kathmandu's Soul Lost in a Concrete Jungle." "People think that KMC [the municipal authority] acted extremely quickly in order to build the garden in Maitighar," the article read. "But they should not forget that KMC had the plan ready and was waiting only for the green signal from the government. Otherwise, how could we have developed such large *mandala* paintings overnight?" (*Spotlight* 21 [2002]:41). This statement highlighted a general uncertainty over who, precisely, was driving the dramatic environmental changes, not just at Maitighar but all over the city. In general, the changes were effected by "the government," but that government was newly opaque. What was certain was that the king had assumed control; precisely who else held and exercised power was noticeably blurred.

As with Tinkune, Maitighar's location was highly visible for SAARC dignitaries being ferried between the airport and city center; it was also considered an urban space that had developed informally. Like the Tinkune case, once Maitighar demolitions and park construction were announced, the ten households there were given fifteen days' notice to vacate and were offered compensation at rates below market value.

During its construction, a *Kathmandu Post* article reported that the government was "working on a war footing" to build the sixty-four-square-foot mandala in the southwest corner of the plot, a stupa in the

Figure 4. The finished mandala at Maitighar, 2005. *(Photo by the author)*

north, and a set of three *dhunge dhara*s (water spouts) in the southeast. Among these features, the massive nested circles of the mandala seemed to command the most attention, if not curiosity. Constructed of a huge iron frame set in the ground, its design rendered in brightly colored rocks, the mandala gave the park construction an air of "making the traditional *mandala* in religious occasions" (Manandhar 2001:1). One could not observe this sculpture without recalling the importance of the mandala diagram to historical representations of the monarchy and the monarch's power. Indeed, it invoked historical patterns of an ancient Kathmandu Valley, developed in a mandala pattern such that social groups were organized according to concentric zones that concentrated power at the center (see chapter 1). In this moment, when the very survival of the state of Nepal as my informants knew it was under such explicit threat, a symbolic rendering of this deeply resonant symbol could not be divorced from the monarch who had garnered his power and imposed the emergency. Along with the act of making it, the very built form of Maitighar could be read as a restatement of the centrality of Kathmandu, the national capital and the seat of a still-reigning monarch, in a larger Nepal gripped by revolutionary violence.

The erasure almost overnight of unplanned urban space, and its rapid replacement being an iconic symbol of the relationship between the monarch and the national project, transformed Maitighar into a spatial rendering of nationhood at a time when the kingship was under undeniable strain. Much more so than the larger park at Tinkune, the new park at Maitighar was an unveiled reminder of historical connections between Kathmandu's moral geography and the legitimacy of the state.[8] A friend assured me that the mandala was visible from the air and would be seen by SAARC dignitaries just before their flights landed at Tribhuvan International Airport.

Military personnel protected the space and reinforced the complications of gaining access to this island of land surrounded on all sides by heavily trafficked roads. Nearly impossible to reach on foot without swift and creative street-crossing skills, I saw no one actually use the park during the emergency; people only gazed from the periphery at the enormous iron and stone mandala forged into the ground.[9]

Although I would not be able to fully appreciate the power of his statement until much later, in 2001, Huta Ram Baidya urged me to take seriously the contest already being fought between the kingship and alternative state imaginaries at Maitighar. "The mandala . . . is a waste," he said. "Just watch it. The stones in the iron frame aren't colorfast, and in a few monsoons the mandala itself will be washed out." The mandala's resilience, like that of the kingship, would soon be brought into active question here in environmental territory. By 2005, Baidya's words proved startlingly prescient.

But let us consider first yet another feature of beautification enacted in the immediate aftermath of the emergency declaration. The SAARC preparation project that intersected most directly with my field research took place in early December 2001. After a routine crossing of the Bagmati Bridge between Kathmandu and Patan, I was overcome by a sense that something fundamental to my naturalized sense of the riverscape below had changed. Within a few moments I realized that the entire sukumbasi settlement previously standing at the base of the bridge was gone. Any evidence of the settlement had been completely removed—erased so fully that a first-time passerby would know it only as a solid dirt road bordered on one side by a trickle of Bagmati River flow.

I slowly reasoned through what had happened. I was aware that the community had been served a notice to vacate a month earlier, but I also knew that like all past notices, it was received as routine harassment rather than as an actual statement of intent to evict. The emergency and general sense of political flux had disturbed much that was routine, however, and the assumption that this notice would go unenforced, like so many others, proved dangerously wrong. Rather than continue home, I headed to the office of the housing advocacy organization with which I worked closely at that time, to gather what information I could about the eviction.

At the office, I found workers sitting outside, gathered around a rattan table in the winter sun. They confirmed that the settlement had been razed, officially due to SAARC-related "security preparations."[10] In response, the group was documenting what had happened and helping to coordinate relief for displaced families, but no public expression of dissent or formal protest was planned. Even in the context of the emergency, this struck me as extraordinary, both because tactics to raise public awareness were the core of the organization's operations and because this was the first eviction of its kind in Kathmandu since 1996. It was perhaps even more unexpected because the previous few years had seen numerous official gestures toward securing the lives and settlements of Kathmandu's sukumbasi.[11] In many ways, activists working in the housing sector had grown confident, reassured that forced evictions were no longer a threat to the growing population of squatters in Kathmandu.

Unlike in the cases of Tinkune and Maitighar, the evictions of settlers below the Bagmati Bridge did not involve any form of compensation for the destruction of property. In all, twenty-two families from settlements on both riverbanks were displaced in two phases, on December 4, 2001, and on January 4, 2002. A follow-up NGO report noted that most of the evicted families were immediately forced to erect temporary shelters on private land near their former homes. Thus, the evictions simply displaced squatters to within a few kilometers of their previous settlements (Lumanti Support Group for Shelter 2002a).

"SAARC summit security" measures such as this one resonated in complex ways with urban elite sentiments and anxieties that accompanied the extensive growth of informal settlements in Kathmandu, particularly along the Bagmati and Bishnumati rivers. By razing some of the most visible among these settlements, the emergency government immediately ascribed order to environmental spaces perceived to be chaotic, spaces that also served as material reminders of the asymmetry of development and wealth distribution in Nepal. Along with increasingly becoming reminders of the political disorder in the countryside as well, the settlements were a logical focus for a government seeking to display its capacity to control and curb migration patterns often framed in popular discourse as a rural invasion of urban space.[12]

The evictions at Thapathali marked a critical shift in official posture toward sukumbasi, and by extension toward rural-to-urban migration itself. This particular SAARC-related gesture not only swiftly eliminated a set of informal urban settlements, but it also communicated an official intolerance for the particular kind of urban fluidity signified by growing rural-to-urban migration.[13]

WELCOMING A POSTDEMOCRATIC URBAN ECOLOGY

In discussions with informants, I expected expressions of outrage, caution, or at least dissatisfaction over the undemocratic context within which SAARC-related environmental improvements took place. Many river-concerned actors had spent the previous decade expressing passionate commitments to ecological visions that took concepts like participation and democracy as central features; I expected them to interpret emergency environmental management as a searing violation of these principles. But to the contrary, and in sharp contrast to reactions to the emergency that I encountered among friends and my host family, many river-engaged actors offered largely positive assessments.

The fact that urban environmental cleanup associated with SAARC preparations had an immediate connection to the image that the city would project to regional and global onlookers figured into this enthusiasm. A pride in simply achieving a change in the environmental space was reflected, for example, in comments made by the Nepali director of the UN Park Project, which had become a symbol of the river restoration stalemate because it remained unbuilt.[14] "Kathmandu is the capital. We want to make the city clean for our guests. These parks were a special case because we saw what can happen when we focus our expertise and concentrate our resources. Everyone worked together and something got done. Finally, some action — it is a relief."[15]

The Thapathali settlement eviction removed what the director considered a major obstacle to his own efforts to move forward with UN Park construction. On several occasions, he said that international donors would not get involved with the UN Park until the "sukumbasi problem" was resolved. Only in the context of a state of emergency, he told me, could such a swift removal of the settlements take place. Thus, this created his sense of "relief" despite the authoritarian nature of the emergency.

Again, contrary to my expectations, even some housing advocates welcomed the swift environmental changes. After the Thapathali actions — the first forced sukumbasi evictions along the riverbanks since 1996 — I encountered more approval than criticisms for the state's actions among these informants. One prominent activist for urban housing and environmental rights expressed relief that, as she put it, at least "something" in the way of urban environmental improvement was finally "getting done," in sharp contrast to the ineffective democratic decade. She emphasized a broader context in which all SAARC-related improvements were a welcome relief, suggesting that, despite their vulnerability, even the city's sukumbasi population had a generally positive reaction to the beautification campaign. After all, she said, sukumbasi were "just like us": Even they welcomed actions that would evidence a proactive, effective government and promote a cleaner city. "People [in sukumbasi settlements] are not against [the SAARC-related projects]," she said. "They say they want these development programs. They say they want the city to look clean. The message the people are getting is that the government can do it if they want... even these big houses [at Maitighar] are demolished in a moment. [The government] can do it if they want to."[16] While her previous work strongly suggested that she would never approve of forced evictions, her reaction to these particular actions framed them as a welcome sign of life for a state otherwise feared to be too unstable to "do it if they want to." In this sense, any action that proved the state was organized provided evidence of vitality and it inspired hope for state survival in the civil war. The only other imagined alternative, a Maoist Nepal, would bring consequences too uncertain and feared among my urban informants for them to support. Thus, this activist went to great lengths to explain that in a broad sense, authoritarian river cleanup measures were in this case acceptable, even to the very communities that could potentially fall victim to evictions.

To interpret such comments as antidemocractic would be inaccurate, however. In fact, informants often invoked an imagined, but as yet unrealized, experience of democracy and ecology rather than Nepal's prajatantra when they explained why they would temporarily tolerate the emergency. Recall that 1990s democratization was riddled with excesses, and it slowly became associated more with cultural and environmental degradation than with positive change. To maintain the possibility of a future, truly viable, and functional democracy, some believed that the integrity of the nation-state had to be preserved first. The contours of an alternative Maoist state seemed either too unknown or threatening to accommodate this aspiration for more "authentic" democracy. If the emergency government could regain control and order, the work of cultivating a truly new Nepal could begin.

For some, then, salvaging the state by backgrounding dissent during the initial state of emergency also meant preserving the possibility that a more desirable democracy—and urban riverscape—might be forged at some future point. The unspoken trade-off was that if the authoritarian practices of the early emergency continued, hope for a more acceptable democracy would have vanished with, to some extent, their own consent.

DEMOCRACY AS DEGRADATION

My informants' sense of relief was in part a response to the ways that urban environmental degradation and dysfunctional democracy were considered to be interconnected. As Kathmandu's situation worsened through the 1990s, public discontent with the gross imbalance between environmental reports, development investments, and official promises on one hand, and steady environmental decline on the other, intensified. The capital witnessed a proliferation of citizens' groups and NGOs concerned specifically with the urban environment, with some devoted solely to the cause of stewardship of the Bagmati and Bishnumati rivers.[17] Figures such as Huta Ram Baidya promoted his idea of the "Bagmati Civilization" (*bāgmatī sabhyatā*) via radio, television, and print media, while the prominent industrialist Binod Chaudhary issued a high-visibility, open appeal to the then prime minister, Girija Prasad Koirala, to either clean up the Bagmati within six months or he would do it himself.[18] In 2001, a group of media and nonprofit professionals promoted the first of what would become an annual Bagmati rafting event, featuring well-known entertainers and media figures braving the brown rapids of the monsoon season to draw attention to the cause of river cleanup. Through their demands for wider public awareness of, and attention to, the accelerating decline of the city's rivers, these groups exposed and criticized the ineffectiveness of successive democratic governments and a development apparatus that, despite years of studies, policies, and projects, had failed to curb river decline.

The *Kathmandu Post* captured popular and professional frustration in its headline "Billions Spent but the Rivers Still Polluted," noting that over three billion rupees in investments toward improving the Bagmati and Bishnumati rivers had been spent but with no discernible positive result.[19] Although a host of environmental protection laws were established during the 1990s, most lacked effective enforcement mechanisms or clear delineations of bureaucratic responsibility. Several government agencies were assigned the responsibility of protecting and managing the urban environment, but confusion about precisely which ministries should manage what tasks often resulted in cross-agency conflicts rather than actual enforcement or project implementation.

A few days after the "Billions Spent" article was published, an editorial response appeared in the same paper under the title "Protect Valley Rivers." Its author argued that ultimately only "concrete measures" undertaken by a strong and effective state could reverse urban river damage:

> There are over a dozen authorities supposed to be concerned with improving the cleanliness of rivers. But lack of coordination and accountability problems mar their performance. . . . In this backdrop, it can be concluded that serious action must be taken without delay to enforce relevant laws and regulations to improve the plight of the Valley's rivers. Unless the government comes forward with concrete measures, rivers that constitute the foundation of the Valley's old civilization will not stand any chance of survival, forget about improving their condition.[20]

In this framing of river degradation, administrative failure had both immediate environmental implications and the deeper, more far-reaching effect of neglecting history and cultural heritage. Bureaucratic ineffectiveness and weakness, then, risked not only poor public health or water quality, but it also held in the balance the heritage narratives through which Kathmandu made cultural and historical sense as the capital city of the Nepali nation-state.

The idea that river degradation was culturally insulting circulated in various media; among the most prevalent were the articles, interviews, speeches, and special television coverage that featured Huta Ram Baidya. Embedded in anxieties about the cultural losses signaled by a degraded river- and templescape were concerns that democratization encouraged excessive individualism and excessive greed. A 1996 *Kathmandu Post* article entitled "How Holy? Much Sewage Has Run Down the Bagmati" by A. Mainali profiled stories of personal offense that interviewees associated with river degradation and related them to an emergent "self-centered attitude" in the Valley:

> Recalling from the past, she could not suppress the ecstasy in her eyes when she referred to the purity of the water of the Bagmati just a decade ago. "We used to bathe in the river and wash our clothes. The water was so clean that people even used to drink. The self-centered attitudes of the people, together with the deterioration in religious values, have converted the sacred river of the Hindus into an abominable gutter, making a mockery of their own cultural beliefs," Padma Kumari said.

Later in this piece, the "self-centered attitude" is linked to the reinstatement of democracy, in which, "like birds released from their cages, [citizens] harbored the misconception that they could do practically anything in a free world." The article goes on to outline recent assessments of water quality, a lack of coordination among various governmental regulatory bodies, and the absence of an overall master plan for riverscape restoration as being key obstacles to reversing river degradation.

Later, on April 5, 2000, a *Kathmandu Post* editorial entitled "My House Beneath Bagmati Bridge" offered a direct critique of post-andolan democracy through an expression of dismay over uncontrolled river encroachment and the corruption that seemed to facilitate it.

> This is democracy. Money is the only thing we have left to call our culture. A handful of money can hush them for sure. . . . You can claim any land in this lustrous city provided you know some officials personally. . . . So long as the government falls upon the hands of those who never care whether the Bagmati came first or we did, more and more houses will be built along the ever-narrowing banks of the poor Bagmati. . . . And if you really have power or money in your pocket, join us. It's far more lucrative than crying out for stopping the pollution in the river.

Democracy, critics argued, created conditions wherein the pursuit of personal financial gain automatically superseded a broader responsibility to steward environmental and cultural resources. Without some evidence that democratic governments not only honored a wider spectrum of historical and cultural responsibilities but were capable of implementing change accordingly, river restoration of any kind seemed impossible.

In interviews with river activists about a perceived relationship between democratization and river decline, certain aspects of degradation were particularly prevalent. Among these, riverbank encroachment by elites and extremely poor rural-to-urban migrants was interpreted as evidence of ineffective law enforcement, and therefore an ineffective state.

In late December 2001, I walked along the Bagmati's banks with Huta Ram Baidya and three others: an NGO worker, a writer for a national newspaper, and a photojournalist. The three shared middle-class status but were from different ethnic groups; their families had lived in the Valley for their entire lives, and so they considered themselves to be "from" Kathmandu. Together, they had founded a small environmental NGO that focused on raising awareness about the city's rivers.

During the walk, members of the group commented that democracy had worsened Bagmati degradation. When I asked if they saw democracy and urban environmental degradation as being directly linked, my question received a unanimous "yes." I asked how, and they explained that democracy brought large numbers of "outsiders" to the city. Rather than referencing the migrant poor, the term "outsiders" here marked members of Parliament and those in their employ. The journalist explained to me that MPs were notorious for their greed and for what she called their lack of "respect" for the riverscape. Together, we looked across the river at a large mansion that had been recently constructed, illegally, in the floodplain by a high government official. The home was evidence of river encroachment by democracy's elites; such homes were an increasingly common sight that testified to an abuse of wealth, power, and privilege that to these young activists partly characterized democracy itself.

The group further suggested that, contrary to popular conceptions, more blame for river degradation rested with elite riverbank encroachers than with the riparian sukumbasi poor. Complicating the more common narrative of migration (discussed further in the next chapter), this description of democracy and environmental decline focused on the bureaucratic elite. The individual lifestyles of those who administered democratic rule were seen as a driving force behind river encroachment, and therefore river degradation.

The very idea of encroachment suggests parameters for who, precisely, may claim riverscape space, and this also connected to a sense of democratic government failure. Again, referencing certain sectors of the Nepali news media is instructive. For example, in "I Need to Get Back My Bagmati," in the *Kathmandu Post* in 1999, Anu Lohani describes a visit to Pashupatinath that relates government irresponsibility and ineffectiveness to a general breakdown of the cultural order at this historical and contemporary site of Hindu ritual practice and reproduction of political order in Nepal (e.g., Sharma 2002). Lohani was "shocked to see the rapid growth of [riparian] settlements . . . built along the sides of the sanctified shrine. . . . Is it something that is being allowed by the government?" the author asks.[21]

Such letter writers associated the rivers' condition with a general lack of state leadership under democracy and lamented a loss of Hindu territorial order. They expressed deep frustration over the obstacles to ritual activity created by river degradation or the colonization of riparian areas by new settlements. In addition to criticizing the government for its practical failures, then, these writers appealed to a specific national and religious imaginary of appropriate river use and order. It was this order that prajatantra, through river degradation, was said to violate.

The parks that quickly replaced disorderly and ecologically degraded urban areas gave physical testimony to the perceived failures of the previous decade of democracy. By showing what an emergency state could "do in a moment," these actions also reminded river-concerned actors how few successful interventions had been undertaken in twelve years of prajatantra. Urban beautifications embodied in preparations for SAARC resonated locally, then, as the converse of government-as-inaction; in this sense, urban park construction and clearing informal settlements signaled a postdemocratic state that was aggressive and effective. The combination of the SAARC preparations and the emergency thus allowed the state to portray itself as a foil to failed democracy and to divert attention from the open question of the sustainability of its own actions and methods.

Herein lies an important aspect of informants' "welcome" of emergency ecology. While the dissent that might have been openly expressed during the 1990s did not disappear, it was measured, reserved, or muted so that the postdemocratic state could be observed and even condoned, the state apparatus symbolically reinforced, and the immediate perceived threats associated with the People's War curtailed. Support for an effective, assertive state, particularly in a time of suspended civil rights, seemed to suggest that democratic environmental management replaced by authoritarian environmental management was still acceptably "environmental." Even if urban parks were constructed without public consultation or accountability, they were nevertheless built.

Embracing the undemocratic practices of authoritarian rule on the basis of environmental action was in this sense ironically consistent with an ultimately democratic imaginary. But let us also notice the role of political uncertainty: For urban river-focused actors, the only apparatus through which they had imagined and engaged river change, both in their past and into their future, was that of the existing state-development order (see chapter 2). Officials, NGO activists, and development professionals all depended on that very order to provide the symbolic and material structure of their encounters, their projects, and their modes of forging environmental change. A willful trade of urban environmental failures for emergency ecology also points to their position within the state-development apparatus itself.

AT THE MANDALA'S CENTER

Although it lasted much longer than my informants expected, King Gyanendra's absolute rule eventually ended. In June 2006, five years after the emergency was imposed, I traveled to Kathmandu for a period of summer fieldwork. I found the city gripped by a new excitement. In April of that year, massive demonstrations and strikes throughout Nepal had successfully forced King Gyanendra to reinstate Parliament. An unprecedented alliance between the Maoists and Nepal's mainstream political parties had demanded that the king yield his absolute control, and massive demonstrations followed. The protests were soon called the "People's Movement II" (jana andolan II).

When Parliament reconvened, it voted unanimously to curtail the monarch's control, beginning the process that would eventually strip King Gyanendra of all power. In May 2006, peace talks between the Maoists and the government reconvened for the first time in three years, and the residents of the capital found new reasons to hope that a transformed Nepal could finally witness functional democracy and the end of civil war.

The monarch's future was cast in serious doubt, and the unmaking of administrative and symbolic elements of the royal kingdom began. Previously ubiquitous photos of the king and his family vanished from public view; longstanding "royal" entities from the Royal Nepal Army to Royal Nepal Airlines were renamed overnight, striking out the "royal." Throughout the summer, news items carefully tracked parliamentary efforts to acquire a full and detailed accounting of royal assets held in Nepal and abroad; in everyday conversations friends

Figure 5. A sack covers a statue of Prithvi Narayan Shah, Nepal's "Great Unifier," outside Singha Durbar in Kathmandu. The statue was damaged during the April 2006 demonstrations against King Gyanendra's emergency rule. *(Photo by the author)*

speculated as to whether King Gyanendra would ever consent to the nationalization, taxation, or even surrender of his family's vast holdings of royal property. Deep anger toward, and opposition to, the king and his family—particularly his unpopular son, the heir Paras—were expressed openly, publicly, and with boldly transgressive performances of defiance.[22] At the same time, friends and informants quietly wondered whether and how King Gyanendra might respond to the unprecedented public insults. Some expected another violent crackdown, which the king could effect by capitalizing on the enduring loyalty of an army that was still believed to be under his control.

After the April 2006 demonstrations, the possibility of new forms of democracy, differentiated from the 1990s experience primarily by the term *loktantra*, which conveyed the idea of an inclusive democracy and a dramatic change to monarchical power, dominated political discourse.[23] For months, Kathmandu witnessed continuous, large-scale demonstrations by myriad previously marginalized groups demanding a say in the content of the interim constitution that would guide the remaking of Nepal's democratic structure and constitution. Groups came to the city to claim inclusion in the nascent "new Nepal" vision and apparatus, amplifying their appeals in daily strikes, traffic disruptions, and protests at key government offices.

The park at Maitighar figured prominently in these appeals, bringing the urban greenspace mandala of the 2001 emergency into the production of yet another, this time democratic, new Nepal.[24] On my first visit to the park after returning to Kathmandu in June, I recalled Huta Ram Baidya's prediction five years earlier. The colored stones that composed the mandala's design had indeed faded—in part because *people* now occupied the park.

Among the hundreds of groups that demonstrated at Maitighar that summer, one caught the attention of the entire city. In July, bonded laborers from five districts in the mid- and far-western Terai regions of Nepal camped in the center of the Maitighar mandala. Officially "freed" from indentured conditions akin to slavery by government decree in 2002, these predominantly Tharu laborers, referred to as *kamaiya*, had traveled to the capital from their home districts to voice their demands to the new government.[25] News services reported conflicting figures, but the contours of the story were the same: between thirty thousand and thirty-six thousand kamaiyas[26] were formally "liberated" from labor bondage in 2002, but the majority remained landless and subject to threats of displacement. Most held government-issued identity cards but were denied the material support to which they were entitled under the 2002 emancipation law.

The demonstrators demanded that the government honor its pledge to provide them with land, building material, and cash (5 *katthas* of land, 35 cubic meters of timber for building shelter, and NRS 10,000, according to the emancipation law). During the day, they took their demands to the land reform office, the parliamentary complex at Singha Durbar, or into traffic in the parliamentary vicinity; at night and in the mornings, they cooked, ate, and slept at Maitighar, in the center of the mandala.[27]

After a week of protests, hundreds of kamaiya demonstrators were arrested and released, and charges of brutal police tactics, particularly toward kamaiya women, were prevalent in the elite mainstream press. Media attention to this group, who in terms of history, caste, class, and a host of other indicators could be argued to be among Nepal's most critically marginalized peoples, prompted broader debates about the freedoms that could be imagined or rendered possible in a new Nepal. Before April 2006, it would have seemed extremely unusual for such disempowered groups to come to the capital in these numbers, and it was even more unusual that they would be so widely acknowledged.

Just after the protests came to a close, an editorial in the *Kathmandu Post* asked, "Are the Kamaiya Really Liberated?" Khagendra Sharma wrote, "Now the ex-kamaiya are agitating for their proper justice. Even without their justice they have contributed their might in the popular movement for the restoration of democracy. They are part of the sovereign people whose power has put the present government in its proper place."[28] The occupation of the mandala at Maitighar underscored this new claim to sovereignty. The park location made the demonstrators visible to commuters throughout the city, but it also placed the kamaiya in the center of a symbol that had itself once stood for monarchical power.

The dramatic story of kamaiya freedom granted in 2002—but still unrealized in 2006—resonated deeply with an urban public still reeling from the struggle against authoritarian rule and swept up in the question of a national future without a monarchy. Kamaiya demonstrators in the middle of the mandala voiced demands for inclusion among a "sovereign people" whose own future was in that moment being forged anew.

There is much to understand about kamaiya participation in the demonstrations of 2006 that cannot be captured in a snapshot of mandala occupation. But the place of protest underscores the importance of urban environmental sites as stages for articulating the "state of the state" to multiple audiences, and this shows that they may never be the exclusive domain of state power. Maitighar was as potent a site to perform an authoritarian state's strength as it was to convey wholly new visions of the state that Nepal could become.

SOVEREIGNTY CONCENTRATED AND DIFFUSED

In 2001, emergency environmental transformations did more than appropriately welcome visiting regional dignitaries or signal a new political order centered on King Gyanendra. They also provided city residents with spectacles of tangible state action in the wake of a long and frustrating legacy of environmental inaction. Each environmental project portrayed an emergency state that was the converse of its prajatantra legacy. Rather than fuel dissent, this initially—but only temporarily—garnered the support of the urban environmental activists and professionals whose previous political actions and statements conveyed deep commitments to participatory democracy.

Disillusionment with the government's capacity to manage Kathmandu's environmental problems after jana andolan I made urban greenspaces a logical place for the emergency state to distinguish itself from its democratic predecessor. This was accomplished in part by the same beautifications that conveyed a particular kind of modernity, vitality, and regional membership to dignitaries and onlookers visiting the city for the SAARC meetings. Authoritarian beautifications were widely embraced as a "relief" for the capital city, even among activists strongly committed to participatory governance, and even as the park's stated audience was limited to SAARC.

But that sense of relief could not be extracted from a political moment when the bureaucratic apparatus that had, since the 1950s, evolved in complex connection to the international development industry now faced the threat of its revolutionary undoing. This left nearly all of the remaining river advocates with whom I worked to forge a logic of complicity and support, in part because no matter how undemocratic it was in that moment, it was nevertheless a form of governance and change making in which they were themselves ultimately embedded.

But their consent was also temporary and itself contingent on the broader possibility of forging a new order more consistent with the river actors' stated aspiration—democracy. That possibility hinged on an alliance between the Maoists and the so-called mainstream political parties, which diffused urban fears of an imagined rural revolution and, once operationalized, proved no match for emergency rule. That sovereignty had diffused through the political events that coalesced in jana andolan II was evident, again, in urban environmental territory.

Although its stones were not colorfast, the Maitighar mandala's resilience as an index of politics was clear. Kamaiya—and, indeed, much of the Nepali nation left out of the modern development period—made their demands of a state yet to come precisely here, in the very spaces reclaimed for old symbols of social hierarchy only a few years before. From their position at the park's center, those from Nepal's furthest margins conveyed a new national vision through a logic of rights. There, in the center of the mandala, they performed a natural and national order of things that boldly detached the mandala from monarchy and also separated the monarchy from the nation.

5
Ecologies of Invasion

ALTHOUGH EMERGENCY BEAUTIFICATION projects like the park at Maitighar later became important sites for claiming a place in the new Nepal, the communities that occupied these same spaces before the emergency remain unexamined. Long before the kamaiya voiced their demands from the city's newest mandala, other marginalized groups were targeted and removed in the name of state security and environmental improvement. This chapter returns to the sukumbasi population along the Bagmati and Bishnumati riverbanks to explore the bureaucratic processes that produced urban *de*population—in particular, the role of urban ecology in turning informal populations into "matter out of place"[1] with no claim to urban ecology. It then engages the shift in which that group would later become a vanguard of environmental improvement in the new, post-emergency Nepal.

As we have seen thus far, human actors often refract assessments of ecological vitality through moral logics (Worster 1994) and specific subjectivities (Agrawal 2005). In matters of environmental quality and housing, these points of view often shape debates about public health or vulnerability to disaster so that they reflect broader concerns about "proper" socioecological relationships to the landscape. At the same time, as in the case of Maitighar, the moral logics of ecology are not predetermined; they may be as useful to, and utilized by, marginalized groups as they are by those holding power.

This chapter traces changes in official and popular representations of Kathmandu's riparian sukumbasi population as the political situation in the capital grew increasingly unstable. Recall from chapter 3 that rural-to-urban migrant settlements along the riverbanks grew dramatically in the decade before the emergency. At that time, ever more visible and seemingly simultaneous processes of riparian settlement growth on one hand, and intensifying urban river degradation on the other, reinforced popular belief in a causal relationship between the settlements and the river's degradation. Advocates for housing rights vigorously contested this, but, as this chapter elaborates, their appeals to "sustainable human settlements" reached a point at which bureaucratic conditions for listening were no longer in place.

I describe how officials, policymakers, and activists debated who was, and who wasn't, "in place" through ecological assessments of the riparian zone. I focus on the logics through which those actors then collapsed social difference, generated or changed the meaning of social categories, and, in doing so, sought to stabilize the "unruly" boundaries (Smart 2001) between built environments considered to be formal and those that were informal. How were inhabitants of informal riparian shelter implicated in shifting diagnoses of urban ecological disorder? How did official actors use these diagnoses to frame migrant settlements as invasive in a manner transgressive of both river territory and moral geography (Creswell 2005)?[2] In certain moments, urban ecology practices situated informal settlements in a broader narrative of degradation, but years later, the residents' informal status would help to legitimize their place in the city's first public "eco-friendly" housing resettlement scheme.

In this analysis, I concentrate on the voices of activists and NGO workers who claimed to speak on riparian migrants' behalf or to act in their interest. By focusing on activists' and NGO workers' place in an official sphere of development and environmental policymaking, we can appreciate how official urban ecological knowledge is made and remade, and we can trace how new categories of degraded river and migrant were mutually produced, and then effaced, as sukumbasi were dissolved into an undifferentiated category of insurgents. We can then look beyond the emergency to understand how a new category, that of the ecologically noble migrant, could eventually form from a legacy of discourses in which migrants were synonymous with disorder.

The dynamic relationship between informal housing and the urban environment in Kathmandu cannot be isolated from the broader dynamics of the "crisis of cities" first discussed in chapter 1. Deepening anxiety over potential environmental catastrophe in the cities of the global South is an undeniable feature of twenty-first-century urban and environmental policymaking, planning, and scholarship. Emphasis on both the magnitude and unprecedented nature of twenty-first-century urbanization constructs a dire problem of growing cities in which matters of housing quality and formality are often engaged through ecological logics and categories like Mike Davis's "slum ecology."

The very concept of informal housing, and the related notion of urban informality, has a long genealogy. Scholars have traced the contemporary emergence of the idea of an "informal sector" to the 1970s; by late in that decade Caroline Moser, writing in *World Development*, defined the informal sector as "the urban poor, or as the people living in slums or squatter settlements." AlSayyad (2003) notes the fact that "even though 'informal sector' embodies a broad set of activities and people without clearly identifiable characteristics, scholars continued to represent it by means of a dualistic framework" (10).[3] That dualistic framework categorizes some communities and activities as "formal" and others as "informal": these designations inevitably obscure deep and sometimes interdependent relationships between the two. While the concepts of formality and informality and its attendant problems have deep traction in international development policy, practice, and anxiety over the future form of growing cities, its meaning must be continually reproduced through the social processes I review in this chapter.

First, I begin by exploring narrative constructions of migrants as river degraders in the mid- and late 1990s, prior to the emergency of 2001. Second, I discuss changes in ecological assessments of migration and slum housing in the wake of the emergency. Finally, I explore a post-emergency slum relocation scheme that resulted in Kathmandu's first "ecofriendly" slum resettlement in 2006.

RIVER "DEGRADERS" IN THE DEMOCRATIC DECADE

In my earliest inquiries into perceived relationships between the growth of urban riparian settlements and urban river degradation, I was often encouraged to question the legitimacy of migrant claims to landlessness and to avoid sympathy for individual migrants or their communities. When a language teacher learned about my research plans, for instance, she joked that, in addition to the word *sukumbāsī*, which is the term generally applied to rural-to-urban migrants in

Kathmandu, I should also learn the word *hukumbāsī*, as this, she said, was what most sukumbasi really were. Whereas sukumbasi means someone who has nothing, the prefix *hukum-* indicates someone who wields power. Implied in her suggestion was the notion that, although they appeared to be powerless, the sukumbasi population was actually in full control of its territorial destiny.

State and development officials involved with river improvement often characterized riparian sukumbasi communities in a way consistent with my teacher's "hukumbasi" logic. They questioned the authenticity of sukumbasi landlessness claims, and they implied that occupying riparian land was more often a tactic to benefit from possible resettlement grants than the result of poverty or desperation. As I reviewed development and policy literature related to urban river restoration, I found this sentiment reinforced in nominally environmental discussions as well. Consider, for instance, this excerpt from the Bishnumati Corridor Improvement Project, a major restoration undertaking on the Bishnumati River: "Most of the squatters living in the Bishnumati Corridor are not . . . bona fide landless urban poor, but instead are merely land grabbers or those in their employ. If existing settlements are legalized, or at least seen to be through upgrading, it is likely that the rate of squatter growth will increase dramatically" (HMG/ADB, 1991).

I occasionally heard a further delegitimizing characterization of the urban riparian landless: they were, many claimed, of overwhelmingly Indian, rather than Nepali, origin. This use of the category of "Indian" fell in line with a range of outsider qualities perceived to threaten Nepali national sovereignty and security.

Images of "land-grabbing hukumbasi," of foreign origins or with foreign loyalties, can be traced in part to a history of state-encouraged internal migration and its unintended consequences. While the details of this history warrant a separate discussion, it is important to note that the historical emergence of a sukumbasi population in Nepal is in part a consequence of historical state- and nation-building efforts. Scholars of migration and resettlement in Nepal have meticulously detailed how discriminatory state population resettlement and land-grant schemes in rural Nepal actually produced new forms of landlessness and squatting, particularly in the 1950s and 1960s (e.g., Shrestha

1990; Shrestha and Kaplan 1982). Specific resettlement schemes catalyzed internal migration at a scale that the state was ultimately unable to control, and many migrants were denied promised land grants once they reached their destination (usually in the southern Terai region of the country). This not only created a population of sukumbasi, but it also was a lasting popular and official conflation of state-sponsored internal migration initiatives and uncontrolled encroachment, land speculation, and fraud. In contemporary conversations about riparian squatting in Kathmandu, state and development informants often invoked these sentiments as the "lessons" of past frontier land-grant projects. Likewise, housing rights advocates cited this history to explain state suspicion of the legitimacy and motives of sukumbasi and to explain state resistance to land distributions or resettlement proposals.[4]

Official ideas of river restoration intersected with images of migrants in specific ways. In an interview with a high-level Asian Development Bank official involved with restoration efforts on the Bishnumati River, I raised the question of his "ideal" vision of restoration.[5] Without hesitation, he responded, "The rivers should be lined with parks, restored temples, and, most importantly, high-end housing." In other words, the riverbanks should be made the most expensive, and most desirable, places to live, not the least desirable and least expensive, as they were at that time. The reality of the present, in which the riverbanks harbored the city's poorest communities (which he repeatedly called "eyesores"), was the opposite of the development ideal, and it was a clear violation of this official's sense of appropriate class territorialities in a modern, developed city in proper ecological order.

These logics of class and legitimacy were supported by an ecological diagnosis in which riparian "land" settled and claimed by sukumbasi was characterized as the riverbed. According to development planning documents by Stanley International et al. (1994), severe channelization from sand harvesting and reduced flow from municipal outtakes upstream had prevented rivers from flowing at their "previous levels" for many years. Restoration schemes called for resubmerging exposed sand flats through an elaborate system of weir dams. These would trap sediment during the annual monsoon season and raise riverbed levels. Such a restoration prescription placed many sukumbasi settlements squarely in the river—on riverbed that, in a restoration scenario, would be resubmerged by river flow. Sukumbasi could thus be taken as nothing less than obstacles to restoring that flow, having claimed river territory as land in a way that was inconsistent with ecological order. Their land claims were thus rendered as illegitimate in urban environmental terms as well as in legal terms.

This idea of ecological illegitimacy was corroborated by popular and official perceptions that accelerated informal settlement growth had caused river deterioration. Migrants' proximity to the rivers gradually naturalized them as an assumed cause and aspect of river degradation. Although nearly all of Kathmandu's sewage flowed untreated into the river system, for instance, riparian migrants were often disproportionately implicated as the cause of declining water quality. This description from the Bagmati Basin Management Strategy reinforces a conflation of insecure tenure and incapacity for environmental stewardship: "Because sukumbasi have illegally settled, they feel insecure and therefore care little for the riverine environment which they occupy. Few of their houses have toilets or proper solid waste disposal services, and their wastes flow directly to the rivers" (Stanley International et al. 1994:A3).

If riparian migrants were indeed incapable of "caring" for the rivers, their out-of-placeness could only be remedied through removal and relocation. But embedded in the problem of whether and how to resettle riparian migrant communities were competing logics of restoration. Just as ecologically driven narratives linked migrants and river degradation, alternative, ecologically driven counternarratives sought to de-link them. Recall from chapter 3, for example, the ways that housing activists used "sustainable human habitats" and "healthy cities" to challenge dominant portrayals of slum settlements as automatically degrading and environmentally harmful.

When I asked if popular suspicions of sukumbasi claims to landlessness weakened activists' "sustainable human settlement" logics, the director of Lumanti, a major housing activist NGO, replied:

> The sukumbasi are not hukumbasi—they are not rich people, although probably some of them own something somewhere—a small piece of land in the village, a small house somewhere, probably that is true.... But [the fact that] they are here in the city [tells us that] they don't have any earning opportuni-

ties there. The legitimacy of the landless? This question has to be looked at as how do we provide affordable housing for the poor?... How do we address this now, and for the future? Poverty is shifting from rural to urban. How do we make the city ecologically sustainable?"[6]

For this activist, ecological improvement hinged on a broad understanding of the social, economic, and environmental processes that brought the sukumbasi population to the riverbank in the first place. To evict them from their settlements would not make this population of urban poor vanish; it would simply disperse them to a new location where, presumably, they would again be considered to be in some way environmentally harmful. Ecological improvement, in her framing, could only truly be achieved when the problem of adequate, affordable shelter was addressed in tandem with the biophysical predicament of Bagmati and Bishnumati degradation. This would require comprehensive and accurate data, as well as a logic of housing as a basic environmental and human right.

Toward the former, Lumanti regularly assisted scholars collecting demographic data, including extensive household-level data collected by Tanaka (1997) that demonstrated vast caste and ethnic diversity in the settlements. A later study, by Hunt (2001), reinforced Tanaka's findings and showed that the primary motivation for migration among sukumbasi was a lack of sufficient land or employment in their place of origin.[7]

SETTLEMENTS IN A STATE OF EMERGENCY

The struggle over whether sukumbasi were ecologically in or out of place on a restored riverscape assumed new dimensions in the context of the emergency, declared in late 2001, nearly six months after most of Nepal's royal family was murdered under questionable circumstances. Debates about the poverty and legitimacy of the landless were at this point officially and popularly reframed in terms of the rural-based, revolutionary political movement that was now widely regarded as a serious threat to the state. Migrants fleeing rural violence were portrayed as the catalyst for a new kind of crisis in the capital. The His Majesty's Government of Nepal (HMG) National Habitat Committee reported that "in the last five years, urban population has grown tremendously not only because of its natural growth and city light attraction, but security and safety in rural areas. Unless the crisis is solved, the movement of rural people to the cities does not seem likely to stop. Consequently, heavy pressure on already scarce basic services such as water, sanitation, electricity, etc., has been created" (2001:26). Slowly, previous narratives of ecocultural degradation or riverbed invasion took on more overt political inflections as riparian communities were suspected of harboring political dissenters or security threats.

This invasion was inflected with a particular tension: on one hand, more and more migrants were acknowledged to be refugees *fleeing* brutal violence in the countryside; on the other, riparian settlements represented a relatively uncontrolled space where the rural dissent symbolized and articulated by the rise of the Maobadi might assemble and take refuge in the city. "Eyesore settlements" on the Bagmati and Bishnumati riverscape thus testified daily to the troubling consequences of uneven development, including their most overtly violent ones, which threatened to entirely rework the sociopolitical landscape through which all environmental change took place.

I resumed a period of fieldwork in Kathmandu a few months before the 2001 emergency was imposed. At that time, despite the uncertainty that followed in the wake of the royal massacre in June, I found housing advocates at Lumanti extraordinarily optimistic about the future. There had been no forced sukumbasi evictions in the city since 1996, and, particularly in 2000 and 2001, the Kathmandu municipality had been conducting what housing advocates interpreted as a promising dialogue about the upgrading and legalization of some riparian settlements. Municipal representatives attended Lumanti-sponsored rallies and awareness-raising events, and they even cooperated in an effort to issue identification cards to sukumbasi families. These were all thought to be extremely encouraging signs. The ideas the organization espoused, of "sustainable human settlements" and "healthy cities," seemed to be taking hold in official policy. Activists hoped for a near future when policy would attend to urban poverty with the same resolve with which it approached environmental improvement. But the rising expectations that accompanied this apparent revision of official views of urban ecology were brought to an abrupt halt with the declaration of emergency in late November.

The most explicit official shift took the form of forced evictions. These happened almost immediately and without public recourse because of emergency restrictions on public expressions of dissent. As violent incidents such as bombings and abductions became more frequent in the capital, so, too, did violent raids on sukumbasi settlements that officials and media labeled as "security checks."[8] Riparian settlements became the default location of risk to the state in its urban capital territory, an assumed automatic refuge for rebels.

During this period, the rhetoric used by many state and development officials about river restoration changed. The settlements went from being a source of annoyance, mere "eyesores," to becoming openly and explicitly identified as *the* primary obstacle to restoration. In interviews in late 2001 with the director of the UN riverside Park Project (first mentioned in the introduction), I was assured that the sukumbasi settlements located inside proposed park boundaries were not only a "nuisance" but were the single most important factor limiting international donor interest and involvement in (and therefore the progress of) the UN Park Project. The director Laxman Shrestha told me, "This land is not clean for donors," arguing that it was, now more than ever, the government's responsibility to "manage" sukumbasi.

A few weeks after this interview, the Thapathali settlements to which the director referred were forcibly evicted. State officials defended the action, citing a "security concern." That same afternoon, the director and I met again for a previously scheduled meeting, and he discussed the evictions with elation and relief, saying that the government had been "freed to act" by the emergency. He said, "These settlements must be brought under control, for the good of the environment and for the survival of the city." He then smiled and told me that, at last, ecological restoration of the Bagmati and Bishnumati could be realized; it was no longer a distant aspiration.

In January 2002, the Asian Development Bank revived a long-dormant initiative to improve environmental conditions along the Bishnumati Corridor. A central element of this project was a road, the Bishnumati Link Road, which was intended to improve traffic flow and relieve congestion in central areas of the capital. Since much of its length was planned alongside the Bishnumati, the road required the removal of existing riparian sukumbasi homes.[9] Officials advised affected families that they would be compensated for the value of their lost homes, so housing advocates pursued municipal and government authorities to produce a written compensation agreement. Officials and activists resolved that residents whose sukumbasi status could be verified as "genuine," that is, genuinely landless, would be paid NRS 2,000 (approximately U.S. $27) monthly for a period of three months. By the end of that period, alternative, affordable housing would be provided. Accordingly, some residents of the designated project zone voluntarily demolished their houses, and all remaining structures were bulldozed in April 2002. Compensation, meanwhile, was delivered unevenly,[10] and questions of how and where to resettle affected families were left unanswered.

Sukumbasi identification cards and other previous causes for optimism lost much of their potency in this period; they could no longer be interpreted as symbols of possibility or evidence of state commitments. Advocates went from promoting healthy cities and habitats to struggling to maintain any voice at all in a debate increasingly dominated by concerns for state (and city) defense. A Lumanti volunteer described the changed terms under which the organization could engage existing debates in February 2002:

> Preventative action and delaying tactics like mass demonstrations, press conferences, people blocking the bulldozers—. . . things Lumanti has done for past eviction threats—can't be done because the possibility of the squatters and staff getting arrested is too high, and the emergency makes people very afraid that if they go to jail, they won't get out until it's over. Second, Lumanti folks feel like they can't single out government representatives as targets for action because they're afraid of being called terrorists. . . . On top of this, if the elections really aren't held this year, then the municipalities are going to grind to a halt.[11]

Short of documenting the situation, previously optimistic activists were virtually paralyzed.

On June 22, 2002, about seven months after imposing the state of emergency, King Gyanendra abolished local governments to further consolidate his power. This left Kathmandu's mayorship vacant for nearly three months, and housing advocates did not have a clear sense of who decided municipal policy. When a government secretary was eventually appointed to the post, he refused to honor prior written agreements related to the Bishnumati Corridor, declaring instead that alternative housing would not be provided. Meanwhile, the demolished homesites of those who were once promised legitimate resettlements remained vacant and undeveloped.

RELOCATING SETTLEMENTS IN SUSTAINABLE FUTURES

In the post-emergency period of dramatic transition from monarchy to republic, the fate of the capital city's vulnerable migrant population again resurfaced in official and public discourse. When I returned to Kathmandu for a period of follow-up research in the summer of 2006, I found colleagues at Lumanti again optimistic, having recently experienced what they considered to be a favorable conclusion to the displacements related to the Bishnumati Corridor Environmental Improvement Project. In the aftermath of the emergency, collaborative efforts between housing advocates and municipal officials had restarted, and eventually a resettlement site was identified and acquired. The financial terms of rehousing were said to be extremely favorable, allowing families to draw low-interest home loans from an Urban Community Support Fund administered by Lumanti.[12] This was not simply a resettlement scheme, then; it was also an effort to extend homeownership opportunities to untitled, displaced migrants.

Activists' enthusiasm about the resettlement site was driven as much by ideas of environmental improvement as it was by the financial incentives that promised to turn displaced migrants into future homeowners. The new housing was unique because it was, according to a Lumanti promotional pamphlet, a "precedent setting," "eco-friendly" housing development.[13] The very migrants whose presence on the Bishnumati riverbanks was once perceived as environmentally degrading, and, later, politically dangerous, now assumed citywide prominence for pioneering ecologically sensitive urban living.

Designed in consultation with the displaced families who would eventually settle there, the new housing site featured large open spaces, a rainwater-harvesting apparatus, and an on-site wastewater treatment and graywater reuse scheme. These elements combined to form a showcase of ecological order.

I joined Lumanti advocates for a much anticipated site visit, eager to hear from residents about their new homes, and to see the overall design of the development. My enthusiasm waned as our van inched its way through, and then beyond, Kathmandu's congested streets, far from the urban core, into Kirtipur—and then on to Kirtipur's outskirts. This, a housing advocate assured me, was the closest site that could be feasibly acquired, and the uncertain implications of its significant distance from the urban center for inhabitants' livelihoods were a necessary trade-off.

As our van rounded a bend and began to descend over a rolling dirt road, the housing development emerged—tucked between the road we traversed and steep, emerald green terraced fields. Lumanti workers guided me through the site, pointing out the open spaces, rainwater-harvesting apparatus, and wastewater treatment facilities. I was given a comprehensive tour of some of the homes, and some residents gathered in the common courtyard. Lumanti workers asked them how they felt about their new homes, and their replies followed a notably consistent script of contentment, relief, and gratitude for the assistance of the government and housing advocates.

But how might one make deeper sense of this slum ecology "success" through the relocation of riparian squatters in an ecofriendly town on the outskirts of Kirtipur? In order to understand my formal tour of the settlement I would obviously require a more grounded, longer-term inquiry into the recent lived experiences of resettled residents—an inquiry not possible at the time, and at this writing as yet to be done. However, the very fact of the resettlement site, the foregrounding of its ecologically sensitive characteristics, and its place in a broader story of Kathmandu's housing and environmental politics, underlined the dynamic nature of what urban ecology means in relation to political change.

The geographic and ecological shifts evident in the relocation were striking. As noted above, migrants once marginalized as environmental degraders assumed potential citywide prominence as Kathmandu's vanguard of sustainability—but now they were on the margins of the city and the margins of everyday urban visibility. On one hand, the Kirtipur site was more periphery than city, disconnected from the economic opportunities associated with the city center. But on the other, through relocation and re-placement, previously landless families might now realistically aspire to landownership, that condition assumed to precede responsible environmental stewardship according to developmentalist assumptions consistent with ecological modernization theory (Fisher and Freudenburg 2001; York and Rose 2003). Ownership would not be of "conventional" homes but rather of buildings with structural characteristics that implicitly challenged previous assumptions that landless migrants shared an ecological pathology. Or did it reinforce them?

First noted in chapter 3, Tiwari's "rural in the urban" had become, it appeared, simultaneously valorized and re-ruralized in a way that suggested that the green rehabilitation of informality could only take place outside the city. I immediately recalled historical urban organization in the Kathmandu Valley, which relegated the lowest castes to the furthest outskirts,[14] but it was also clear that the simultaneous shift out of the city and into formalized green homeownership collapsed economic and ecological moralities. Further striking was the way that re-placing riparian migrants seemed to simultaneously achieve a riverscape that was "clean" for donors and development (according to Shrestha, the UN Park director), and, as Michael Herzfeld noted in his study of the evacuation of urban market areas, the relegation of "potentially 'dangerous' populations to spaces where they can be subjected to increased surveillance, and away from those spaces where their continuing presence is indeed viewed by the authorities as 'matter out of place'" (2006:132).

It is important to notice that the convergence point for ecological, developmentalist, and economic moralities was in this case the built form—housing. In this case, ecofriendly dwellings seemed to promise simultaneous environmental and social reform. The legitimizing power of environmentally sensitive resettlement was immediately clear; it seamlessly merged with Lumanti's longstanding commitment to a forward-looking vision of urban ecology that focused first on housing rights for the poor. Yet it also reinforced a perceived need to reform sukumbasi housing practices even as it challenged old stereotypes of slum dwellers incapable of caring for the environment. It also transported the entire community to a place remote from the urban core, a domain they might imagine re-entering only after having assumed formal landowning status.

The social and environmental rehabilitation associated with ecofriendly sukumbasi housing reworked the relationship between the form of housing that migrants occupied and the kinds of urban citizen-subjects they might become. Here, ecological sensitivity reinforced an expected bridge between informal and formal subjectivity, conditioning a path from the city's margins to its core that was consistent with the moral logics of a particular urban ecology. Migrants were not simply resettled, then, but replaced—in political, moral, and environmental terms—through interventions derived from intersecting problematics of housing and environment.

INFORMAL HOUSING AND ECOLOGIES OF REFORM

The relocation of Bishnumati Corridor migrants raises questions about the political work that discursive and practical intersections of housing provision and sustainability perform. Ecology was a fundamental concern when defining the built form of a future Kathmandu as well as the relationship between that built form and the moral practices of its inhabitants. A Kathmandu developed in a more sustainable guise promised to reform an ecologically problematic set of housing practices and sites, as well as, perhaps, its new or untitled inhabitants.

In this way, urban ecology in practice framed a discursive and material politics of place and belonging—belonging to the river system and to the core areas of the capital city. In the 1990s, settlements of rural-to-urban migrants were an increasingly prominent, visible reminder of the spatial inequities of decades of national socioeconomic development in Nepal, and as such they stood as reminders of state failure situated, visibly, in the state's own bureaucratic and symbolic heart, its capital. This was Tiwari's "rural in the urban"—the rural brought into the logic of the capital city, and the city's response in the form of anxieties over mounting environmental pollution, degradation, and decline.

But in the context of the emergency, this "rural in the urban" came to stand as well for the political disorder of Nepal's rural revolution—recast through the emergency as a legitimate threat to the state. In this political context, urban environmental interventions became more explicit, overt gestures of state control—maneuvers to be understood in the broader arena of war.

When development initiatives resumed under the post-emergency democracy called *loktantra*, ecofriendly resettlement produced a spectacle of reform, replacing, and in the process remaking, an entire community and its place in the city. The environmental sensitivity of resettlement housing reinforced the idea of squatters as invaders and degraders in need of reform, while situating them at the vanguard of a new kind of urban poor, whose housing was more sustainable and, would become, in the long term, both economically and ecologically formal. Only then, presumably, might they re-enter core urban space in a way deemed legitimate and consistent with eco-developmentalist logic. In the meantime, they could also showcase the transformative potential of a new Nepal, organized through new state and socioenvironmental imaginaries.

It is not insignificant that riparian areas along the Bishnumati were resettled as part of a major riverside road-building initiative, hardly a facet of urban development that has automatic ecological benefits, but it is one with clear links to developmentalist modernization. Recall as well Huta Ram Baidya's observation, recounted in the introduction, that so-called formal housing on the riverscape presented its own socioenvironmental problems. In 2006 it was still unclear whether the form of housing actually permitted along the Bishnumati River would itself be subjected to ecological scrutiny.

Even as its moral inflection shifted, official ecology-in-practice presupposed a single, homogeneous category, sukumbasi, throughout the period reviewed in this chapter. This category automatically collapsed considerable differences, including the presence of a full range of caste groups and dozens of ethnic groups among the riparian migrant population. Action research studies highlighted this tremendous diversity, beginning with Tanaka's groundbreaking (1997) sukumbasi demographic profile. However, the official sphere required a category that would signify all informal housing practices and mark them as singular, regardless of the caste, class, or ethnic identities contained therein. It is indeed the case that in this sense a modern ideology of urban renewal and environmentalism supplanted longstanding cultural norms regarding status and appropriate territoriality along the riverscape.[15]

Caste, class, and regional differences among riparian sukumbasi gained new relevance to the state in the period of emergency, when state agents feared that settlements could harbor Maoist insurgents. Yet the category that captured all settlers and settlements prevented selective state intervention in those sukumbasi communities thought to pose a specific and demonstrable political risk. One consequence described here was that all migrant communities received automatic state suspicion, and some were subjected to direct violence. This demonstrates quite precisely the material implications of official ecology-in-practice and its power to collapse social difference.

To contextualize more conventional notions of a global urban ecological crisis is to raise important questions about how the urban environment is problematized in rapid growth cities and about how institutions at a variety of scales respond. A long legacy of critical environmental scholarship has focused on how the environment is "made" through formal efforts to save, conserve, restore, or protect it (cf. Greenough 2003, Grove 1989, Sivaramakrishnan 1999). Yet in rapid-growth urban contexts like Kathmandu, those same spaces of restoration may also be sites of settlement, complex zones of struggle over precisely which territory constitutes human habitat and which constitutes the urban environment. While the processes of delineation, infusion with meaning, and constant negotiation are fundamental aspects of how nature is made and remade in any setting (cf. Zerner 2003), the twenty-first-century urban context introduces new and unique stresses, contests, and calculations about just how much room for nature exists in cities. And there remains, in its urban guise, Raymond Williams's (1980) enduring question of the ideological work that ideas of nature perform.

6
Local Rivers, Global Reaches

IN CHAPTER 4, I noted that the river-focused actors with whom I worked were all, to some extent, embedded in the complex state-development apparatus against which they often struggled. By virtue of their professional participation in river restoration, they affiliated in some way with institutions linked to transnational circulations — monetary inputs, discursive regimes, and sometimes transnational mobility.[1] In most cases, the livelihood of these advocates depended on these circulations. While this did not make significant change to Nepal's state-development structure impossible to imagine, it sometimes conditioned initial responses to official actions like the emergency.

These same actors, however, regularly positioned themselves as locally grounded in the cultural, historical, and political context of modern Kathmandu. In a manner similar to what Appadurai (1996) has called "the production of locality," their river-focused work wove together the analytical domains we might call global and local in the practice of their profession, even as they constantly negotiated their positions in each.

This chapter looks more carefully at the dynamics of position by considering two debates over Kathmandu's riverscape that had clear and explicit extralocal links. Rather than imposing prefigured categories for global and local on these cases, I am interested here in the ways that river-focused identity and global connections were strategically invoked or rejected. I consider the ways that the performance of local and extralocal ties conferred legitimacy to certain constructions of ecological meaning, toward understanding how urban ecologies and their moralities gain traction at multiple scales.

A rich body of scholarship has addressed the perils of analyses that presuppose fixed, polar analytical categories called "global" and "local."[2] Hannerz (1996), Tsing (2000, 2005), Sivaramakrishnan and Agrawal (2003), and others have suggested alternative categories, such as "habitats of meaning," "multiple global ties," or "regional modernities," respectively. Sivaramakrishnan and Agrawal (2003:13) remind us that "both global and local are to be understood in terms of the relations they signify." These authors define localities as "produced as nodes in the flows of people and ideas . . . [they] are thoroughly socially constructed" (ibid.:12). As for the global, they argue that "in a very real sense there is no global—since there is always rupture, disjuncture, and variation in what is imagined as the global" (ibid.).

Following this analytic, this chapter reviews two encounters in which actors drew on global, regional, and local discursive resources to construct, and strategically convey, the specific "meaning" of urban river ecology. First, I describe a debate about riverscape monuments to demonstrate how the competing frames of urban ecology first discussed in chapter 2 could be used to embrace or critique the Nepali nation-state's relationship to global bodies like the United Nations and international development institutions. Second, I describe the dynamics of global and local positionality in an encounter between two river awareness-raising groups. Here, the challenge of forging local connections considered critical to effecting river change became a platform for debating grounded collective capacity to improve the riverscape.

Let us begin with the place of two pillars—one imagined and one extant—in river activists' aspirations for restoration.

REMEMBERING NEPALI SERVICE

The UN Park, as previously described, was a riparian environmental improvement project proposed in July 1995 by a group of acting and former state ministers, influential scholars, and development professionals. The park's official purpose was to commemorate the fiftieth anniversary of the United Nations. In its earliest formulation, the project plan called for greenspace to replace sandy flats of exposed riverbed on the banks of the Bagmati and Bishnumati. It was a bold proposal to dramatically increase the public, multiple-use park spaces in the city, but its ecological trade-offs included the fact that it would also make permanent the morphological change associated with river degradation. Advocates claimed that such trade-offs were well worth the cost, emphasizing the park's tremendous symbolic potential, derived in large part from the fact that it would "mark respect for those Nepalese army, police, and civilians who have died in UN peacekeeping endeavors."[3] The project was linked to the centrality of the capital city in the broader national landscape and to the nation's place in the modern twentieth-century global networks that the UN represents. In its early stages, park boundaries fluctuated, but by 1999 a master plan covered river corridors along much of the Bagmati and Bishnumati in Kathmandu.[4]

Despite its symbolic global connections and anticipated appeal among Kathmandu's middle class and elite, the park stood unimplemented and largely unbuilt. Its only physical manifestations over the course of my fieldwork were cemented edges along the Bagmati River in the proposed project zone and the sukumbasi evictions undertaken in the Teku-Thapatali area (described in chapter 4).

Operating from inside the Ministry of Housing and Physical Planning complex, the UN Park Project office employed a director and, periodically, active teams of consultants. Global festivities marking the fiftieth anniversary of the United Nations came and went in the mid-1990s, but an apparent lack of interested donors left Kathmandu's UN Park vision unfunded and unimplemented.[5] It remained a widely discussed proposal, however; if funded, the park promised to reshape and restore the high-visibility Teku-Thapatali riverscape in very specific ways.[6]

The park plan changed over time, usually in response to perceptions of donor interest. Facilities ranging from children's play and picnic areas to parking lots and other public services were included in most versions, but some features appeared and disappeared. For a brief period, the master plan included an interactive exhibition called Mini Nepal. Intended to showcase the ethnic and cultural diversity of Nepal, Mini Nepal would contain what the UN Park director at that time called a "typical" house, staffed with people wearing ethnic dress and preparing foods associated with that group, for every officially recognized major ethnic group in the country. The director explained with great enthusiasm that such a theme park would bestow a variety of benefits; among these was the opportunity to educate the population of the Kathmandu Valley about their fellow national citizens from other regions. "I myself would love to visit such a park," he told me in an interview. "I would like to see a real Sherpa house, and a real Tharu house."⁷ The education that the director imagined urbanites could glean in Mini Nepal, then, was cultural and national — a performance of the makeup of the nation, and its national project of forging unity-in-diversity, laid out in recreational riparian greenspace at the center of the capital. The director Laxman Shrestha, a Kathmandu Valley Newar, noted that "especially in times like these" (referencing the political upheaval of the People's War and a withering kingship), a project like Mini Nepal could "remind" Valley residents of the social diversity beyond the capital and the enduring national need for Nepali unity. Unmentioned but implied was the parallel reminder achieved by locating such a spectacle of constructed diversity on Kathmandu's riverscape. Already sounded at Maitighar, the concept of Mini Nepal underlined the historical and contemporary centrality of the city; it was, and would continue to be, the locus of national bureaucratic power, national identity, and, in the present, modern "green" progress. Situating Mini Nepal in the heart of the Kathmandu Valley, and constructing it under the auspices of a UN commemorative park, reinforced the spatiality of the nation; in this sense it had clear utility for contemporary Nepali state making. The park in general, and specific features like Mini Nepal, carried the potential to restate the legitimacy of the ruling regime and its predecessors who had entered into membership in the United Nations in the first place.

For Shrestha, such symbolism was positive: "As the center of Nepal," he said, "we should have a natural place we can go to experience the whole country." If that center could become a green recreational space that resonated with contemporary global legitimacy, he told me, the rivers' overall ecological well-being could be recognized, rescued, and ultimately safeguarded. His concern, he said, was for the health of the Bagmati and Bishnumati rivers, and for him the connection between river revival and the symbolic appeal of the UN Park was almost automatic.

Like many other aspects of the project, however, Mini Nepal was abandoned when it failed to attract donor interest. In fact, the director considered only one element of the park design to be absolutely indispensable: a memorial pillar. Since the park was intended to affirm

Nepal's membership in the United Nations, the pillar would mark the place to honor the Nepali soldiers who had served and died in worldwide UN peacekeeping missions. In so doing, the pillar and the park would also stand for Nepal's contemporary global significance and long legacy of what Shrestha called its "international bravery." I was struck by the pillar's indispensability and the urban ecological outcomes the director believed it would produce.

During an interview in the winter of 2002, Shrestha suggested that we visit the proposed park area. From behind a hospital near the Bagmati Bridge, we walked briskly to the site where he hoped that the memorial pillar would one day be built. After many years of struggling to make the park vision a reality, Shrestha said, he found himself willing to compromise on nearly every detail of the proposed park layout and area—except this one. After carefully describing the pillar's imagined features, he rubbed his hands together for warmth against the winter cold and declared, as though to an audience of many, that the pillar must be built *"Nepālī sebā samjhanako lāgi"* (in memory of Nepali service). It was service to the "international community" signified by the United Nations that Shrestha saw as one of the greatest, most important features of modern Nepal—a marker of Nepaliness and a point of contemporary pride among Kathmandu's middle and upper classes.

There was no more appropriate place to commemorate Nepali UN peacekeepers than the banks of the Bagmati, Shrestha argued, where for centuries Nepal's rulers built temples and other commemorative structures. Then, in a rhetorical moment that brought his hopes for the UN Park into the direct purview of riverscape ecology, Shrestha outlined the potential ecological benefits that he believed would follow even the most modest implementation of the park plan. The memorial pillar, he explained, would make an important contribution to river restoration because it would endow the banks of the Bagmati with *meaning* for the city's modern, globally conscious population. Park greenspace was one thing, but the pillar would attract people there in the first place. Visitors would feel drawn to contemplate Nepal's contribution to the global community and, in turn, the country's contemporary global importance. They would be compelled to remember the rich history of the capital city and, hopefully, its previous ecological state. Triumphant pride animated his gestures as he spoke, giving his sense of the practice of urban ecology in the UN Park an affective resonance with nationalist narratives of past and present Nepali heroism to complement its symbolic statement of global belonging.

Here was Shrestha's sense of ecological utility: by contemplating the message of a UN commemorative pillar, visitors would also contemplate the ecological state of the Bagmati and be inspired to change it. Thus the most important, indeed indispensible, way to cultivate a modern sense of concern for river ecology was to draw Kathmandu's populace to the rivers to experience a connection to past, present, and future collective identities. A pillar honoring Nepal's modern national identity and its place in an international community was, in his view, the most essential part of realizing any biophysical objectives in the UN Park Project, or river management in general. In this sense, the pillar was indirectly — but still fundamentally — ecological, insofar as it would reattract the attention needed to catalyze river stewardship.

After we returned from the proposed park site, Shrestha explained that the pillar had become a personal commitment, to himself "and to Kathmandu." He told me, "Perhaps this UN Park will never receive the financial backing it requires, in which case I am willing to concede defeat. But there is one thing that I cannot abandon. The pillar must be built. Even if I have to build it myself, the pillar must be built. Even if we do nothing else, we can do this."[8] Perhaps the larger project required global funds, but locally — indeed, in profoundly personal terms — Shrestha could catalyze river ecology through a built symbol placed on the river's edge. It would not necessarily require international donors to gather citizens to the site and to encourage them to improve the Bagmati's condition; it would take only the commitment to build a pillar, and in so doing it would provoke contemplation of the pillar's message. Shrestha preferred donor support, of course, since that would facilitate full realization of the park vision, but at its core, restoration relied on local recognition of, and pride in, Nepal's place in the global networks signified by the UN.

Even as Shrestha shared his intention to build the pillar himself, he knew that to enact the UN Park vision more fully, international donors were essential. Whenever the topic of funding arose in our conversations, the tone of his statements turned to anxiety and disappointment. As he explained, Nepal's assistance to other countries —

whether providing peacekeepers to the UN in the present or supplying the British with Gurkha soldiers—was always largely ignored. "This is recognition that does not come easily to Nepal," he said. Even as he sought to commemorate Nepal's place of respect, he conceded that it was seldom fully affirmed in that same global community. He often recounted moments of struggle to simply be noticed by international donors, let alone to have his proposals carefully considered and financially supported.

Participation in, and entitlement to, the particular, powerful global arena signified by the UN was thus central to Shrestha's definition and practice of urban ecology. By encouraging modern uses of the Bagmati while still reproducing particular narratives of its historical and ritual symbolism, the park and pillar connected biophysical river restoration objectives to the ongoing affective work of state making and spatialized narrations of collectivity, affinity, and belonging. He thus anticipated the day that his imagined pillar, and the restored riverbank that would frame it, would come into being.

REMEMBERING DEFIANCE

By contrast, another pillar evoked critiques of the global community associated with the development industry. By calling attention to this pillar, its advocate also promoted a form of river restoration that amplified nationalist history, Nepali heroism, and autonomy, but it also stood for independence from, rather than a proud place within, networks of development power with global origins. While Shrestha aspired to place a new pillar on the Bagmati riverbank, Huta Ram Baidya, the architect of the Bagmati Civilization idea, promoted the proper labeling and treatment of an existing monument, the Bhim Sen Pillar. Baidya's critique of the international donor community and a complicit Nepali state, introduced in chapter 2, contrasted sharply with Shrestha's enthusiasm for Nepal's global connections; his urban ecology had affective resonance with independence and autonomy rather than with global belonging.

In the first few weeks of my acquaintance with Baidya, late in 1997, our conversations focused almost exclusively on his anger about a particular pillar, and its place on what was then a newly constructed bridge over the Bagmati River. The bridge had just been completed with donor assistance from India, replacing a previous structure that was near collapse. The new bridge was larger and sturdier than its predecessor, but one aspect gave it continuity with past architectural styles in the Kathmandu Valley. A large stone pillar, topped with a gleaming brass lion and secured in a stone base carved in the likeness of a tortoise, stood at the bridge's north end. Although majestic, its precise origin and age were impossible for a newcomer to determine, as the pillar itself was unlabeled. Baidya found this anonymity telling and infuriating. Consistent with examples like the Gopal Mandir (chapter 2), he argued that the pillar had been improperly relocated, and then left unlabeled, in a deliberate developmentalist attempt to obscure its origin.

That origin traces to General Bhim Sen Thapa, leader of the Gorkhali Army during the Anglo-Gorkhali War, and later prime minister of Nepal (1806–37).[9] In both roles, Thapa's defiance of British colonial authority is legendary in nationalist discourse. As an army general, he boldly occupied British territories; as prime minister, he maintained a policy of isolation and nonengagement with foreigners of any kind, including the British Resident posted in Kathmandu under the terms of the Treaty of Segauli. Bhim Sen was thus a figure whose defiance inspired pride; his heroism derived not from (overtly) global networks of power but from within the boundaries of Nepali national territory. His brave heroism was associated with his Gorkhali heritage, extended to his Nepali identity.

The Bhim Sen Pillar originally stood on a bridge over the Bagmati that was built during Thapa's rule.[10] Baidya claimed that it was moved to the new Bagmati Bridge in violation of all existing archaeological laws, thus making it "archaeologically displaced." He found it outrageous that, once it was moved, the pillar was stripped of an explanation of its origin. Without a plaque, he said, the pillar's national significance would be lost—refashioned, and perhaps co-opted, by the development powers that now used it to embellish their own symbol of progress: a new bridge. In newspaper editorials and many personal conversations, Baidya called for the pillar to be labeled with its derivation and original location. In a letter to the *Kathmandu Post*, he wrote, "The pedestal deserves this honor. If not done, history will be distorted."[11]

On this and many other river-related issues, Baidya tirelessly challenged the relationship between the Nepali state and the international development community. He emphasized Kathmandu's haphazard contemporary history of urban planning in general, and planning on the rivers specifically, to directly link the dependent relationship between the Nepali state and foreign aid to Bagmati degradation. According to Baidya, Nepal's state-development alliance produced an uneven, distorted relationship between international aid, its experts and planners, and the Nepalese public. He warned that the result would be a cultural and historical amnesia in which development was fully divorced from local historicity and specificity. River degradation, he argued, was a natural consequence of this amnesia, unstoppable in the absence of remembrance and critique.

Baidya's charges hinged, in this case, on a displaced, unattributed, and misremembered (or newly remembered) monument. It carried with it a suspicion of the global institutions and the state's, and therefore Nepali citizens,' relationships to them. Since Baidya saw Kathmandu's river crisis as in part driven by development and its processes, river reclamation would require reviving a particular version of nationalist social memory. This version emphasized a need for independence from donor influences and foreign "experts."

Both Shrestha and Baidya saw their respective pillars as important tools for promoting national narratives that would motivate river stewardship. Indeed, the locally grounded and globalist sentiments that they believed the pillars embodied were critical to conveying the proper meaning of urban ecology. In this sense, both pillars underlined how ideas of nationalism and aspirations for ecological action were tightly intertwined, yet the proper composition of that nation was a clear point of contest. While Baidya sought to strengthen contemporary Nepali national identity, and in turn improve river conditions, by de-linking national history from international institutions and global citizenship, Shrestha sought to reinforce the global attributes of national identity by honoring Nepal's ties to the international community. Neither called for a wholly new state structure; both focused instead on the place of river restoration in contemporary and future ideas of the nation.

While a single pillar cannot eliminate river pollution or remake a river ecosystem, these advocates' passionate endorsements underscore the importance of struggles over narratives conveyed by the form and content of urban ecological spaces. Forging and promoting these narratives is the work of a multilayered state, but it is also the work of those who assume postures of contest against the state apparatus.

Furthermore, just as official framings of river degradation failed to consider global institutional complicity with river destruction (see chapter 2), Shrestha and Baidya seemed far less concerned with conventional biophysical indicators of river ecology than with the symbolic currency of the built riverbank landscape; neither believed that cultural affect and biophysical improvement could be fully disentangled.

BELONGING TO RESTORATION

During the same period, the affective dimensions of urban ecology in practice figured into planning, discussions, and independent advocacy among international development agencies and their representatives as well. By the mid-1990s, Kathmandu's expatriate community, made up largely of development industry workers and their families, was organizing awareness campaigns to promote dialogue and action on the rivers' behalf. In December 2001, I noticed a small advertisement in the English-language *Nepali Times* announcing a meeting of Kathmandu residents concerned about the condition of the Bagmati. The ad listed an e-mail address that one could contact for more information, and I placed a request. A few days later, I received a reply inviting me to meet one of the group's organizers in the courtyard of an exclusive luxury hotel near Baneswor.

This was my first venture inside this hotel. The interior was appointed with meticulously rendered woodcarvings and terra-cotta work in a style associated with Malla Era Newar architecture. All around me, hotel employees in traditional Newar costume waited on guests. The surroundings bespoke powerful connections between specific symbolic representations of Kathmandu and a romanticized, marketable version of the larger country; these converged for the upscale transnational tourists' leisured consumption. They also reminded me of my relative neglect of another population that might mobilize in the name of Bagmati and Bishnumati improvement: international visitors and expatriate Valley residents.

The Bagmati awareness group organizer was a middle-aged non-Nepali woman. Over cups of tea shared in the hotel courtyard, I learned that she had been living and working in the Kathmandu Valley for about a year. Although I did not ask for details, she mentioned leaving behind her own hotel in western Nepal. I assumed that she was in Kathmandu for the same reason as many others who had left the western districts in the recent past: the violence of the People's War and Maoist resistance toward businesses associated with foreigners had likely made it impossible to stay.

As we talked, I learned that the group that had placed the *Nepali Times* ad was called Friends of the Bagmati. It was still in its formative stages, and although it had an extensive membership list that included noted Nepali intellectuals, journalists, and Nepali and expatriate NGO professionals, it had as yet established little in the way of a plan of action or source of financing. Friends of the Bagmati would focus on river awareness-raising activities, but it had settled on few details as to how to create this awareness.

My host explained that the group originated at a conference held in Kathmandu in November 2000, when a British organization called the Alliance for Religion and Conservation (ARC), in partnership with the World Wildlife Fund, convened a conference, called "Journey to Kathmandu: Sacred Gifts for a Living Planet Campaign." Among the attendees was Britain's Prince Philip, who, as part of the conclusion of the conference, launched the formation of the Friends of the Bagmati. The group was born, then, out of international conservation interests and specific interest in the "sacred" quality of the Bagmati[12] by a largely non-Nepali community.

After our meeting, seeking a clearer sense of what motivated this international effort on the Bagmati's behalf, I read the Friends of the Bagmati mission statement. A portion read:

> The Bagmati River is highly sacred to the Hindu peoples of Nepal and neighboring countries. Cremations take place along its banks as it flows past the most holy of Hindu temples, Pashupatinath, and its history is woven into the history of Kathmandu and Nepal. Today, however, ritual bathing is almost a

thing of the past as pollution chokes this once beautiful river. Presently there are no attempts being made to integrate conservation and clean up activities into a wider fresh water management system for the whole valley. It is the Friends' aim, through highlighting the sacredness of the Bagmati, to restore the Bagmati and the other rivers of the Valley by raising local awareness through a variety of ways.[13]

I grew interested in Friends of the Bagmati's strategy. How would transnational activists position themselves in a restoration debate that was already replete with questions of who does, and does not, belong on the restored riverscape? How would it go about raising local awareness, when figures like Huta Ram Baidya had achieved such notoriety precisely by opposing efforts of international origin? How, in short, would a global effort on the Bagmati's behalf operationalize its aspirations locally, and what would be its reception?

My preliminary meeting with the Friends of the Bagmati organizer centered less on the Bagmati's sacred qualities and more on her own uncertainties about what the group could and would accomplish. Rather than emphasize the Sacred Gifts for the Planet Campaign, she focused on local bureaucratic obstacles to effective river management, explaining that the long history of failed attempts to organize Kathmandu's citizenry around river restoration had left most members of Friends of the Bagmati "jaded" and skeptical. Expectations were low, and she added that this was particularly true among the Nepali members of the group.[14]

Drinking tea in the peaceful shaded courtyard, our conversation settled on the organizer's explanation for Nepali members' skepticism. It touched two main themes: a lack of a cooperative spirit in Kathmandu civic life ("no one wants to work together," she told me) and the role of international development institutions and their projects in damaging the city's civic vitality. My host explained that, in her view, huge inflows of development assistance over a long period had cultivated corruption rather than development; she even suggested at one point that all international development aid to Nepal should simply be stopped.

Just a few minutes later, however, I learned that the Friends of the Bagmati aspired to NGO status, and therefore the group had wanted to become part of the very development flows and processes that the organizer had just critiqued. While she was conscious of problematic dimensions of international development, she did not imagine a way that her advocacy group could operate outside the admittedly fraught state-development sphere.

Friends of the Bagmati's first major awareness-raising event was a monsoon river-rafting festival. Together with the Nepal River Conservation Trust and Nepal Tourism Board, it involved thirty-five kayaks floating from Sundarijal over a rain-swollen Bagmati to join with politicians and environmental activists in rubber rafts at Til Ganga. The procession concluded at Shankhamul with a conference and concert. I was not in Kathmandu to witness it, but I read press coverage and discussed it with informants when I returned to Nepal a few months later. Hailed in the press and by some of my informants as a fantastic success, Friends of the Bagmati regarded it as a key catalyst for awareness and river improvement. Under the ironic headline "Brown Water Rafting," Kanak Mani Dixit, the *Nepali Times* editor, wrote for the paper's largely elite English-speaking audience:

> This was not white water—it was the silt-and-sewage laden Bagmati. . . . On the rafts were a host of environmentalists, journalists and some celebrities. Said the satirist Madan Krishna Shrestha, "I always wanted to go rafting, but how could I know the first time would be along the Bagmati and not the Trisuli!" His colleague Hari Bangsha Acharya added, "I can imagine a time when this river will be clear, and we can catch fish in it." . . . As the rafts drifted downstream with their kayak consorts, the Bagmati lost its sand bed in its entirety to the construction industry, and so rather than meander as it used to, the river today flows along muddy clay canyons. All along, untreated sewage joins the river in cascades of dirty fluid. Then there are the poorest of migrant communities who have also colonized nearly the entire stretch. Hundreds of latrines dot the riverside from Tilganga to Sankhamul. The riverside is also used for dumping everything from offal to industrial byproducts, carpet industry sludge, and the generic garbage of the city.[15]

If bringing visibility to the plight of the Bagmati was the group's primary objective, then this event seemed like an effective first step. But how would Friends of the Bagmati achieve its stated goal of involving local people? I wondered how the group would define river locality itself and how media spectacles like brown water rafting would inspire and organize local stewardship.

Some clues emerged a few months later at the next event organized by the Friends of the Bagmati, a riverside "cleanup" at the Kalopul area of Teku.[16] A group of about sixty participants, myself among them, converged at the site, equipped with a tractor, shovels, and brooms on loan from the Kathmandu municipality. We collected trash for a few hours, filling tractor loads with garbage to be hauled away, though it occurred to me that, perhaps ironically, we did not know *to* where the trash was being hauled away. A "Teku Heritage Walk," intended for tourism professionals who would then take what they learned and conduct such walks on their own, was held the same day but with scant attendance.

Few residents of the surrounding neighborhood of Kalopul took part in the main Teku cleanup event. Instead, they watched it, as a spectacle, from the edges of activity. It was certainly unusual, and probably somewhat astonishing, to see foreigners and well-off Nepalis simply appear one morning, in masks and gloves, and spend several hours collecting the many kinds of waste that covered the site. We were almost gleefully undertaking polluting tasks historically associated exclusively with low-caste groups, but they were tasks that in modern, "clean green" ideology were the responsibility of every citizen regardless of social position.

Yet performing modern environmental responsibility had not been the objective of the event. The group's hope, stated in Friends of the Bagmati organizing meetings, was to promote cleanup involvement by local residents, meaning those living proximate to the river. While we attracted press attention and accomplished temporary garbage removal from the Teku area, our goal of involving river-proximate residents had met with utter failure.

The problem of how to promote local participation in events like this was fervently addressed at the group's next meeting, held a month later in the hotel where I'd first met the organizer of the Friends of the Bagmati.[17] As in that encounter, I found myself distracted at first by the luxury of the setting, and its contrast with conditions outside the hotel grounds, to say nothing of the riverscape we were assembled to discuss. Attendees—a mix of foreign resident expatriates, Nepali intellectuals and activists, and short-term tourists—sat around a huge, glossy conference table that spaced us at a gaping distance from one another. As introductory remarks were made and the agenda explained, hotel service staff, dressed as usual in clothing associated with the Newar, served us tea and cookies from elegant silver platters and teapots.

The meeting focused immediately on the challenge of transforming the group's international origins to a practical life as a local advocacy group. One of the main Teku cleanup organizers, a Nepali NGO professional, gave a formal summary of the event, lamenting that it seemed to function more as a spectacle than as an inclusive initiative. She described her personal attempt to encourage local residents to join in and claimed that one line of reasoning to which local residents seemed to respond favorably was that tourism-related benefits might follow from a trash-free Teku area. A few residents of Kalopul were persuaded, she reported, but "taking part" meant doing things like "showing [Friends members] where they could find soap and water to get cleaned up."

The urgent matter of making the Friends of the Bagmati into an initiative driven by local, river-proximate residents (none of whom were present at the meeting, or as far as I could tell, were members of the Friends of the Bagmati) formed the substance of much of the first half of the meeting. The group struggled to devise a way to encourage river stewardship among those living near the river, and in doing so it revealed the presumed association between river proximity and river pollution that I have described in previous chapters. While it would be an exaggeration to suggest that the group regarded river pollution as something that originated exclusively among river-proximate populations, speakers at the meeting made explicit connections between the two. Indeed, it was this very connection that seemed to drive the aspiration to promote local participation in the first place.

The meeting gradually shifted from reviews of the Teku cleanup to the work of defining the Friends of the Bagmati's broader goals. A British expatriate suggested that the group encourage a "neighborhood watch" system on the riverbanks, explaining that a model used in Britain for security purposes could be implemented among residents on the Bagmati for environmental purposes. "We could go in and educate the women who then go and educate the local people," she said. I was immediately reminded of sukumbasi informants with whom I'd spoken who argued repeatedly that their riverbank presence already functioned as a patrol for curbing the riverside solid-waste dumping that was widely practiced by the Kathmandu municipality. In this sense, a neighborhood watch already existed, and it was perhaps the municipal authorities, who continued to dump the city's solid waste on the riverbanks, that needed to be "educated."

Halfway through the meeting, one of the Friends of the Bagmati organizers asked all members to send her an e-mail with a list of the contributions that each individual might make to the organization's activities. At this point, two women who introduced themselves as representatives of an "informal group" called *Hāmro Bāgmatī* (Our Bagmati, hereafter Hamro Bagmati). Their remarks sent a simple but powerful ripple through the room. They introduced themselves in Nepali rather than in the English that had been used exclusively in the meeting to that point, as well as in all other dealings I had had with the Friends of the Bagmati. Proceeding in a language that many present only vaguely understood, but implying that all present *should* understand, one of the women described the origins of Hamro Bagmati and their activities to date, which included meeting with leading Nepali figures in the science and politics of river management. She reported that the group was actively collecting historical and recent river photographs and that they were planning a public photo exhibition.

A freelance photographer and photojournalist, she then shared some of her personal philosophy about photography, relating it to Hamro Bagmati's aspirations to hold a photo exhibition about the past and present Bagmati. She offered an analogy to the Friends of the Bagmati that did not go unnoticed by many in the room when she talked about the distance that often exists between the photographer and her subject. The speaker said that she strove to make her photography a mutual experience between the two. She contrasted this with the approach of a tourist who comes to a place, takes a photo, and leaves, explaining her desire to cultivate genuine connections to the places represented in her pictures.

For those in the room who could understand her comments, it was clear that the woman was drawing a comparison between the Friends of the Bagmati's dilemma of how a global group becomes more local and her own group's interest in breaking down barriers between those who visited the riverscape and those who live there. She knew that not everyone in the room could understand her, but the message implied in her self-presentation was that those who could not understand her could not meet even the most basic criteria for calling the Bagmati *hāmro*, or "ours."[18] By simply shifting to Nepali in a space in which the prevailing assumption was that we were all sufficiently "cosmopolitan" or globally connected to use English, the woman effectively communicated one of her own criteria for effective strategies to catalyze Bagmati restoration. Several members of the Friends of the Bagmati were clearly excluded by this and visibly offended.

The speaker continued for several minutes, but the frustrations in the room soon reached a breaking point, and another Friends organizer, a Nepali professional who had understood the remarks, interrupted. In a scolding tone, she returned to English and said that the ideas that Hamro Bagmati had shared sounded fine, but they struck her as idealistic and akin to "expecting miracles." She then tried to defend the Friends of the Bagmati's position, saying, "We don't have an environment that lets us involve local people instantly, and it's unfair to say they [the residents of Kalopul and Teku] should also be at this table." The Friends of the Bagmati speaker went on to point out that other "equally idealistic" groups like Hamro Bagmati had existed in Kathmandu for years, but thus far they had realized few results.

By this point the exchange lost any pretense of mutual respect and the meeting simply moved on, in English, to the topic of a design for the Friends of the Bagmati brochure. "I want this next brochure to be upscale," one of the organizers told the group, "so I can take it to ARC [the British organization that founded FOB]. That doesn't give it direct funding, but we can use their name to back it up." A group member then announced that March 22 was World Water Day. With this, one of the women from Hamro Bagmati leaned over to me and whispered in Nepali, "Who decided *this*?"

For the Friends of the Bagmati, extralocal connections were fundamental to catalyzing river restoration, and despite a critical awareness of their adverse effects, links to global institutions and systems were in some ways considered essential and unavoidable. As with Laxman Shrestha's UN Park, the Friends of the Bagmati foregrounded the idea that the Bagmati riverscape was globally important and that it therefore made sense to maintain close contact with extralocal material and discursive interests in that riverscape. While just how to cultivate simultaneous global and local functionality was considered a problem, it was one that the group regarded as surmountable so long as local ties were cultivated and global connections maintained.

Hamro Bagmati, by contrast, as it was defined in this particular encounter, emphasized certain kinds of local familiarity and functionality as being central to promoting riverscape improvement. They had no expressed intent to become an NGO or to join those receiving international development funds, and so they did not have to meet global criteria for participating in the discussion (criteria that might demand the use of English, for example). In a manner analogous to Huta Ram Baidya, they emphasized the importance of getting Kathmandu residents themselves to talk and learn about the historical life of the riverscape and to decide how the rivers should be restored. The tension between the two groups at this meeting, and their respective approaches, persisted despite the fact that both shared the ultimate goal of involving local people in the stewardship of the rivers.

Embedded in the Friends of the Bagmati's fervent discussions of how to involve local people was an assumption that "local" would automatically translate to "effective." At the same time, the group's origins in an international arena anchored it to the logics of locality that that arena espoused.

The representatives of Hamro Bagmati agreed that local participation was necessary for effective river restoration. They challenged, however, the assumption that an organizational or individual journey between global and local required nothing more than noble intent. With language as a strong instrument to demonstrate their point, they suggested to the Friends of the Bagmati that actual local involvement, in which river-proximate residents initiate action, required specific forms of belonging.

GLOBAL AND LOCAL ECOLOGIES IN PRACTICE

As biophysical entities, rivers defy any urge to dissect them into discrete reaches, and in some ways they necessitate a conceptualization that transcends the immediate territories a single person or group of people may immediately observe and experience. In this sense, neither locality nor globalism is independently sufficient for ecologically knowing, or managing, a river. The disciplinary urge in ecological thinking is often to consider a system, based on the assumption that biophysical processes are interconnected and organized according to scales that cannot be grasped at exclusively local levels. But when modes of system thinking meet human social practice, the role, place, and contours of locality and belonging become primary determinants of human action. Indeed, they become fundamentally ecological.

River-focused actors' aspirations for the form, content, and meaning of Kathmandu's riverscape drew consciously, strategically, and often explicitly on specific and dynamic combinations of local, regional, and global resources in order to define and enact urban ecology. Thus what urban ecology *meant*—its stakes, its urgency, and its very possibility—derived in part from engagement with extralocal histories, discourses, and resources. The question first posed to me by an NGO director in Patan—"Can you tell me what urban ecology means?"—was clearly central to operationalizing river ecology and to imbuing it with meaning sufficient to motivate river stewardship.

Actors and organizations that invoked or affirmed their global ties linked a sense of modern global citizenship to creating and maintaining contemporary relevance for the rivers among the city's populace. By emphasizing, and physically marking, the riverscape's place in global communities and networks, informants reasoned that Kathmandu's citizenry might value it differently. For Laxman Shrestha, a UN Park and pillar would remind Nepalis that the rivers' importance stood for forms of Nepali heroism and achievement in arenas beyond its geographic confines, perhaps also locating the riverscape in value systems, networks of power, and communities of environmental affinity beyond the scale at which they could be experienced or traversed in everyday life. For Friends of the Bagmati, international ties and donors provided the potential for successful river management, since exclusively local initiatives (like Hamro Bagmati) were regarded as fated to fail.

Those who challenged extralocal involvement in, or control over, river management did so on the grounds that questioned the trade-offs inherent in forging a place of respect in global spheres of power.

These trade-offs marked historical, linguistic, and cultural familiarity, sensitivity, and ownership, aspects of everyday forms of sociality in the river territory that were in many ways at odds with the automatic assumptions of interconnection and affinity suggested by the idea of a common global future. These actors framed global interventions as distorting, co-opting, or drivers of river degradation itself, and their notion of a meaningful and modern riverscape amplified Nepali autonomy from specific forms of regional and global power. Modernity, here, was an axis around which histories and identity formations quite different from those reinforced by their more globalist fellow actors could be cultivated, strengthened, and lived in meaningful ways.

Designating actors' positions as globalist or localist, of course, misses the hybridities evident in each. Even as the Friends of the Bagmati sought to maintain its ties to extralocal resources and support, its primary goal, and self-proclaimed main failure, was catalyzing locally grounded initiative. The group regarded its lack of capacity to do this as a fundamental weakness. Here was an acknowledged need for a complex combination of global and local involvement in river management; the issue was not whether local or extralocal actors were entitled to participate in management of the Bagmati and Bishnumati, but rather to what extent and on whose terms.

Thus emerges the question, "Restoration by whom?" Nepalis' individual and collective potential surfaced repeatedly as individuals ostensibly united by their common ecological objectives debated their capacity to restore the riverscape when constructed as Nepalis, Valley insiders, and global citizens. Even as Shrestha recognized that if they did "nothing else," they could accomplish that, the pillar he desired was just a piece of a larger UN Park undertaking that would require the state-development apparatus and its regional and global resources. Earlier examples like the BASCRP echoed similar limits to capacity. These limits underscored the importance of an effective state, resilient in the face of political challengers and "freed to act" in somewhat surprising ways (chapter 4), so long as they served broader ecological goals.

Conclusion
Anticipating Restoration

BY THE SUMMER OF 2009, Nepal's new state apparatus was taking shape. The monarchy was officially abolished in December 2007, a condition of a peace settlement between the Maoists and a newly convened Parliament. Four months later, elections to a new constituent assembly delivered the largest bloc of seats to the Maoists, and soon after, in May 2008, Nepal was declared a republic. Prachanda, a central figure in the Maoist leadership since the early days of the People's War, became the prime minister, and for a period the "integration" of a revolutionary movement into democratic party politics appeared more straightforward, and even more peaceful, than my Kathmandu-based informants had initially feared. Prachanda's tenure as prime minister was short-lived, however; in May 2009 he resigned for reasons formally attributed to his opposition to President Ram Baran Yadav's firing of the Nepal Army chief. The Valley witnessed fresh demonstrations and strikes, but the postmonarchy parliamentary government, and formal peace between that government and the Maoists, remains in place.

I returned to Kathmandu in June 2009 to attend a celebration with my host family: One of Jyoti's and Krishna's sons would be married during a week of rituals and festivities. It was my first visit since 2007, and I was eager to reconnect with those in my domestic and research spheres. I was curious about the latest condition of the riverscape that I had traced for over a decade.

A brother from my host family met me on my arrival at Tribhuvan Airport, proud to drive me to Satdobato in a car loaned by his employer. As we inched our way through heavy traffic on the city's Ring Road, he told me of grand municipal plans to widen the thoroughfare and to construct another, larger, outer highway. These changes were welcome and needed, he explained; new infrastructure was the only way to accommodate what had become an unmanageable load of private vehicles in the Valley.

In time, we crossed the Bagmati, at a point close to Pashupatinath. I couldn't help but detach from the conversation and gaze out beyond the bridge. "Ah," my host brother said, noticing my distraction. "Let me ask you. The condition of the Bagmati is worse than it has ever been. So many years after you first arrived, it just keeps getting worse. Why?"

Posed in a time of transition from 1990s democratization to a consolidated monarchy, and, eventually, to a nascent "new Nepal" in the making, analytical questions first asked of a degraded riverscape quickly became questions about multiple claims to the future of a capital city. Efforts to ensure, create, or imagine ecological stability were infused with, and shaped by, aspirations for political stability; to promote urban ecology was also to engage in the reproduction or contestation of the cultural idea of the nation-state. In this context, the question "Can you tell me what urban ecology means?" brought the simultaneous making of a modern riverscape and a modern urban future into sharper focus.

It also underlined the ways that, in periods of rapid political transformation, the competing temporalities of state making and ecology often seem wholly incommensurate. While one measures change in increments of immediate urgency, the other stretches that urgency beyond presentist preoccupations to attach its longings for change to long-term cycles and faraway futures.

These issues animate the ongoing methodological and conceptual challenges of studying urban ecology as a socionatural process. Urban ecology's multiple forms and practices in contemporary cities demand an adaptive analytical framework that can assess biophysical change while attending to the production of the social categories, histories, and meanings that legitimize those assessments. Analyses will inevitably capture moments in otherwise dynamic processes, since, as with biophysical change, social categories and histories are never fully fixed and stable. Yet understanding why, after so many fervent efforts, the riverscape was still degraded, demands attention to those moments; through them, we discern the form and content of the claims to moral order through which certain ecological logics were made legible, powerful, and active.

In the case of the Bagmati and Bishnumati, overlapping notions of environmental degradation and state transformation foregrounded important facets of urban ecology in practice. First, they showed how complex and multiple forms of exclusion accompanied particular experiences of the environment and the ways that social groups tried to preserve or re-create certain environmental experiences. Second, since river restoration politics unfolded during different iterations of democracy and monarchy, they illustrated the importance of contextualizing the political orders within which social actors forge their environmental claims.

Third, by considering interconnections between urban ecological and political aspirations, one finds the questions that Ferguson (1999) famously asked of modernity relevant to urban environmental imaginaries. That is, we may usefully ask, what social expectations accompany urban environmental interventions? What kinds of polities do actors imagine and intend when they advocate for particular ecological practices, policies, and outcomes? How do those expectations shape the range of responses that are considered reasonable, acceptable, and moral, and how do those expectations influence metrics of environmental failure and success? These considerations are critical aspects of urban ecology in practice, as they signal social processes that constantly engage, and sometimes rework, the structures within which specific knowledge forms, claims to identity and territory, and tellings of history are acknowledged and legitimized, while others are not.

The Bagmati and Bishnumati case demonstrates the extent to which understanding ecology in practice demands attention to the competing visions of modernity and moral order that fracture imaginaries of an otherwise unified environmentalism. We can claim neither a singular and stable environmentalism nor a singular and uncontested attendant morality. Even as it assembled river-focused actors, the environmentalist goal of urban river restoration was also an arena within which actors continuously constructed, articulated, and practiced differentiated logics of the good and proper modern city that would effect, oversee, and reproduce river vitality.

Those logics gave ecological legitimacy to specific social categories and practices of exclusion. In different ways and moments, actors came to believe, or official practice confirmed, that certain groups possessed the will, reason, and capacity to ecologically transform themselves and the urban spaces they inhabited, while others did not. At the same time, the dynamic character of ecological logics meant that social categories and exclusions were often temporary, malleable, and sometimes completely reversible. In this sense, urban ecology in practice enabled the continuous making and unmaking of social affinity and difference. This study suggests, then, that urban ecologies may increasingly supply and organize discursive and material resources from which differently positioned and differently powerful actors draw to (re)make the state, (re)map urban space, and (re)order urban social life itself. In the twenty-first century, ecology and the environment matter as powerful *social* forces; neither is wholly shaped by biophysical change nor fully separate from it.

In the Kathmandu case, the sociality of urban ecology formed an arena in which new logics of sameness, unity, and belonging were forged; these logics sometimes explicitly cut across historical modes of indexing difference. Expressly environmental affinities represented a possibility for forms of social unity and cohesion that otherwise lacked moral fortitude. This unity was at times considered central to realizing an ecologically sound riverscape fully integrated into a sustainable and desirable political future; environmentalism thus afforded the important possibility of belonging even to the otherwise marginalized kamaiya (chapter 4) or sukumbasi (chapter 5).

In the twenty-first century, multiple forms and logics of ecology in practice signal far more than a set of issues around which temporary, politically active environmentalist collectives form and dissolve; they signal newly powerful ideologies of belonging through which human affinity and difference are socially produced. While urban ecologies organize engagement with modern technologies and biophysical science, they are also sites in which fundamentally social negotiations of otherness, moral order, and collective imaginaries are conducted and fixed into categories and contests.

The Bagmati and Bishnumati case also suggests an enduring place for development studies in analytics of urban social change, even in a twenty-first-century geopolitical landscape riddled with challenges to old mappings of power between the global North and South. Myriad development "failures" characterized Kathmandu's environment in the 1990s, but certain failures animated broader expressions of disaffection with the political order. The plight of the Bagmati occupied a central place in popular itemizations of democracy's disappointments; this remained the case as Nepalis fashioned a new state and social order after April 2006. The environment in general, and the rivers in particular, stood as a ready metric for the "new Nepal's" efficacy and performance, and its relative autonomy from historical relationships with global development institutions and their influence.

This also suggests newly important analytics about the place of nature — specifically urban nature — in contemporary political transformation, and it brings the cities of the twenty-first century into the foreground as they mediate relationships between nature and the nation (Comaroff and Comaroff 2001; Cederlof and Sivaramakrishnan 2006). Environmental expectations and the political imperatives they condition illuminate how regimes of cultural representation can stand for an idealized state (Faure and Siu 1995; Hansen 2001; Kapferer 1988; Mbembe 2001; Sivaramakrishnan 1999) as well as an idealized state-society synergy conveyed through the discourses and practices of urban ecology (chapter 2, chapter 4; Rademacher 2009).

To explore overlapping assessments of political order and natural order is to foreground the ideological work that ideas of nature aspire to perform but never fully accomplish (Williams 1980). It is to consider aspiration, hope, and fear of loss as key aspects of the subjective dynamic of state formation (Aretxaga 2003) and to recognize ecological assessments as both the terrain of rational technologies of management and control (Saberwal 1999; Scott 1998; Sivaramakrishnan 1999) and engagement with those technologies and their organizational regimes. The Kathmandu case shows, in grounded ways, that urban ecological aspirations, fears, and indexes of loss host newly important expressions of the proper, moral organization of space, social memory, and social order (chapter 5).

With the Bagmati and Bishnumati rivers as an ethnographic anchor, I have endeavored to recount a cultural politics of modern, urban, environmental restoration formed in the wake of a democratization process widely regarded as disorderly, destructive, and utterly disappointing. By reviewing points of political and ecological intersection, I have shown how unmet expectations of order through democracy gave way to initial consent for authoritarian river management techniques and the consolidation of monarchical control (chapter 4).

But a discussion of ecology in practice must reach beyond observations about political order and state making if it is to meet its analytical potential. Throughout this book, practices of urban ecology were emancipatory and anticipatory. If we consider Peet's and Watts's idea of liberation ecologies as applicable to the Bagmati and Bishnumati case, we see that urban ecology was emancipatory, as it promised to reorder a riverscape and polity that was constantly marred by perceived disorder. Here, the social potential of urban ecology was formed and reformed in dialogue with competing cultural ideas of the state and with intersecting notions of place and identity.

Urban ecology was also anticipatory. The simultaneous hope for positive change and an engagement with powerful material and discursive tools understood as imbued with the capacity to effect that change drove the processes and contests recounted here. In this sense, practices of urban ecology implied a capacity to reproduce belief in the very possibility of change, that is, to operate socially as facilitators of the capacity to aspire (Appadurai 2004). Riverscape-change narratives were infused with aspirations for broader socioenvironmental transformation, and it was precisely these aspirations, and the belief that they could be realized, that gave those narratives powerful social traction.

Recall the words of Huta Ram Baidya, who regularly accused the state and its official development activities of harming both the rivers and Kathmandu cultural life. Addressing a crowd gathered along the Bagmati River, he shouted, "We have all seen development, which has given us some things. We talk about rights. But what about our cultural rights? Development has destroyed my cultural rights! . . . Our culture has been broken!" It was a formidable charge, that the guardians of Nepali cultural identity—assembled simply as the government— had destroyed their own legitimacy by coevolving with development. This logic of loss then echoed in a very different social position, that of a sukumbasi woman, who told me on the same day that "in development (she had) lost everything." "Our lives," she told me, "are just *sarkhārko khel*" (the government's game). In these accountings of loss, and their performance and pronouncement, actors linked urban ecology to alienation as well as anticipation. While Baidya's idea of Bagmati Civilization entitlement was far from housing advocates' idea of sustainable human settlements or upgraded squatter housing, it was ecology, and the larger aspirations for new social collectivities, affinities, and political orders that it suggested, that anchored these actors' hopes for a better urban future.

That better urban future was itself embedded in concern that rapid urban growth would continue and that its stresses would multiply. In important ways, competing urban ecologies simultaneously explained the present and mapped an alternative to this expected future crisis (chapter 2). Here, the Kathmandu case suggests that practices of urban ecology must be understood as future-making strategies that lend temporality, explanation, and order to as-yet-unrealized forms of anticipated and feared change.

Logics of belonging to the riverscape, and competing ideas of entitlement to belonging, return our study of urban ecology in practice to the social construction of the proper and the improper in the environment and society. Discourses and practices of ecology were infused with moral logics through which ideal types and their opposites were constructed (chapter 5). These ideal types structured representations of the problem while justifying practices and performances of order. The moral power of ecology in practice was its capacity to define the good, balanced, and desirable order of nature *and* society, and that order's opposite—the condition of degradation. But as I have shown here, each urban ecology frame was also inevitably selective and was constantly contested.

In Himalayan studies, a rich tradition of sacred space and landscape literature informs any engagement with issues of moral order and the urban form. In this book I provide an account of the making and unmaking of urban ecology in a present dominated by developmentalist environmentalism and a contingent state. My focus on the dynamic present suggests a need for more work that brings this literature to bear on the impact and importance of contemporary, multiscaled urban processes with forms and logics of environmentalism that are largely associated with globalization, modernization, and technological societies.

In so doing, I have sought to contribute ethnographic accounts

182 Conclusion

Figure 6. The Bagmati River just below the Bagmati Bridge, 2006. (Photo by the author)

of the moral and affective dimensions of twenty-first-century developmentalist urban ecology. Through prescriptions for the good and proper way of stewarding, relating to, and experiencing urban nature and one another, Kathmandu's river-focused actors organized places of belonging, and set the terms through which they regarded themselves and the social body as being, becoming, claiming, or simply hoping for, a desired ecological future.

While the scientific dimensions of ecology give us essential tools to compare across cases of environmental degradation, and to render commensurate and intelligible common environmental conditions in cities around the world, ecology in practice suggests the need to engage the context, and the context-generating power, of urban ecology enacted in place. It emphasizes social knowledge production and its hierarchies, and it suggests that the meanings of urban ecology defy any single, ordered way of knowing, or changing, nature. The Bagmati and Bishnumati case demonstrates that nuanced attention to urban ecologies as sets of social practices situated in time, place, and politics anchors, and in some cases, directly challenges, metanarratives that frame our "urban planet" in aggregate, and assume that it is automatically, and uniformly, mired in crisis conditions of environmental collapse. Experiences of urban ecological change are real and often unprecedented, but the urban nature and social dynamics in which they are experienced in everyday life form mosaics of moral logic, aspiration, and struggles over power. These require us to bring the places and situated practices of urban ecology more fully into focus, and to ask not only how change is occurring, but also when and how dominant environmental narratives sharpen or obscure the full contours of those changes.

But we must not stop at the paralyzing observation that local contexts are complex. Here it is useful, perhaps, to return to our consideration of urban ecology as an "engaged universal" (introduction) in order to foreground the generative *potential* of local encounters and the moments when they illuminate adaptive change strategies that anticipate dynamism and expect disorder. These encounters demand humble and reflexive, and yet perhaps more effective, analytics of urban ecology that retain both global purchase and local effect. Such analytics necessarily abandon delusions of political neutrality or wholly uniform applicability, and instead they seek the best in competing perspectives. They identify consistencies and shared commitments across these perspectives but recognize the significance of power relations and asymmetries.

Appreciating a fragmented and contested, yet engaged, universal called urban ecology reminds us of the absence of neutral spaces from which to imagine or forge environmental change, and this calls on all social actors who are committed to creating more sustainable futures to clarify the political work embedded in our initiatives and interventions. Environmental projects, however neat in their planning and rigorous in their science, must always assume a place in social life, and we must anticipate social change over time. There can be, in short, no theories of ecological change and sustainability without attendant theories of the social orders that they will accompany. Urban ecologies signal, and must capture, both.

Notes

INTRODUCTION

1. *Daś nāmi,* or ten names, refers to the followers of Sankaracharya, an eighth-century Hindu reformer and saint. Construction of this area dates to the time of the reign of Jung Bahadur Rana. The temple was destroyed in an earthquake in 1934 and was rebuilt in 1935. See Amatya (1994:46).

2. There is no unproblematic translation from Nepali to English for sukumbasi, although it is most commonly translated to "squatter." A related word, *sukumbasa,* is the state of having nothing. Used to refer both to people and their settlements, sukumbasi refers to those who are assumed to be landless, or very poor, and who occupy land for which they do not own a legal title. Although technically the term refers to "the person lacking shelter and food; one having neither" according to Pradhan's (2001) *Ratna's Nepali English Dictionary,* some of Kathmandu's sukumbasi population may not be said to be definitely and universally lacking these things. The population is made up of both rural-to-urban migrants and migrants originating within the Valley. In general, residents of Kathmandu tend to refer to anyone "illegally" occupying public land as sukumbasi. The term can carry negative connotations and, although it is widely used, can be taken as an insult. See Tanaka (1997) for more detailed demographic information on Kathmandu's sukumbasi population.

3. Throughout the text, I use the terms "developmentalism" and "developmentalist" to signal the historically varied paths to social betterment promoted by the complex of international development institutions established after the Second World War. Relevant institutions here include the World Bank, the International Monetary Fund, and the United Nations, as well as the governmental and nongovernmental organizations at multiple scales that broadly promote their social improvement agendas and that draw from and contribute to their material and discursive resource flows. Ideologies associated with developmentalism over time include aid-financed industrialization, neoliberal economic and social reforms, specific policy approaches to poverty reduction, and the use of specific metrics to assess and define social progress. My use of "developmentalism" is intended to foreground important connections between an international development apparatus firmly rooted in a capitalist world economy and specific, albeit dynamic, interpretations of the economic, governmental, bureaucratic, and technocratic forms that best embody "progress." I wish to emphasize that developmentalism is always encountered rather than imposed; as such, it may assume multiple forms in the same historical moment.

4. *Kāntipur,* December 21, 2002.

5. See Sharma 2002:30.

6. Pashupatinath's associated deity, Shiva Pashupati, has been worshipped on the Bagmati banks at least since the fourth century A.D., and both deity and temple play an important role in Kathmandu Valley origin myths. Slusser explains that

> to the Nepalese, at least, there is no question respecting Pashupati's origin in the valley and his long association with it. For, as the chroniclers aver, "first of all there was nothing in Nepal except Pashupatinath, whose beginning and end no one can tell." His origin legend, as told in the *Nepala-mahatmya*, also names the Kathmandu Valley as the locale of Shiva's manifestations as Pashupati. Shiva, tiring of the ceaseless adulation of the gods at Benares, thought to masquerade as a gazelle in the Slesmantaka wood of the Nepal valley. The gods traced him thither, and after entreating him to no avail, at length forcibly seized their Lord by the horns, which promptly shattered. Bounding to the right bank of the Bagmati, his present temple site, Siva declared, "since I have dwelt in the Slesmantaka wood in the form of a beast (pasu), therefore throughout the universe my name shall be Pashupati. Visnu then erected a fragment of the broken horn as a linga, and all the gods, including Buddha, hastened to offer their obeisance. In time, the god's temple crumbled, as temples are wont, and buried the divine linga. But at length, a sagacious cow, remembered by many names, sprinkled her milk over the spot, leading a curious herdsman to reveal anew the wondrous linga, Pashupatinatha. (Slusser 1982: 227)

Also, according to Slusser,

> Of the many lingas in Nepal, none has played a more influential role than that which embodies Pashupatinath, Lord of the Animals. One of the innumerable manifestations of Shiva, Pashupati is the supreme god of the Pashupatas, a sect of the ghora type, whose existence in India can be traced to the second century B.C.. The Pashupatas either introduced their deity into Nepal at such an early date that even in India he came to be thought of as a Nepali god, or the Nepali Pashupati in fact represents a local syncretism of Shiva with an indigenous pastoral god, protector of the animals. This is suggested by the nature of his legends, which concern a gazelle and a marvelous cow, and by an allusion in the *Mahabarata*, in which Shiva assumed the guise of a Kirata, an indigene of the Himalaya, to give Arjuna a weapon named Pashupata. (1992:226)

7. This polarity gradually became more of a triad, with three main factions—royalty, political parties, and the Maoists—eventually feuding among one another. For example, see Dhawal Rana, "The Great Game," *Nepali Times* #145, May 16–22, 2003.

8. By "state making," I refer here to analyses of statecraft that foreground the processes through which governments, governmentality, and civil society are formed and legitimated. This approach to the study of the state, exemplified in environmental studies by works such as Sivaramakrishnan (1999) and Agrawal

(2005), considers the production and reproduction of state-society distinctions as objects of study. For a comprehensive discussion of the approach, see Sivaramakrishnan 1999:4–13. For an excellent review of state studies as they relate to Nepal, see Lakier 2005.

9. Lakier (2005) provides a comprehensive discussion of state studies in Nepal and emphasizes the extent to which the cultural idea of the state and its "myths" have been addressed by a variety of scholars.

10. While river degradation and general deterioration of urban environmental quality have accelerated dramatically over the past decade, the processes and discussions surrounding them have much longer genealogies. Two excellent early reviews of popular dialogue surrounding urban environmental deterioration—as well as degradation of the Bagmati and Bishnumati rivers—are "The Valley Chokes" (*Himal* 1:1 [May 1987]) and "Limits to Growth: The Weakening Spirit of Kathmandu Valley" (*Himal* 5:1 [January/February 1992]).

11. For a discussion of Hindu perspectives on rivers, see Alley 2002.

12. Nepali-to-English spellings are reproduced as they appear in the original text.

13. John Metz (2010) provides an extensive discussion of the production of, development processes associated with, and critical responses to the "theory of Himalayan degradation." Julie Guthman's (1997) historiography of the theory suggested that the crisis of mountain degradation in the 1980s and 1990s was a social construction, not in the sense that it had no material basis, but insofar as the knowledge produced to demonstrate it was socially, institutionally, and historically embedded and was deeply power-laden. Guthman delineated three main types of aid regimes since the 1950s: modernization, basic needs, and neoliberalism. In each, she demonstrated how certain environmental ideas became embedded in, and were reproduced through, the political and institutional construction of facts. She argued that the "Himalayan dilemma" (as described by Ives and Messerli 1989) could not be contained as being only an ecological issue; the social and environmental conditions subsumed under the theory were also political, economic, and cultural—as well as scientifically uncertain.

14. This follows Phillip Abrams's (1982) concept of "structuring," which assumes that every social structure has embedded within it a set of processes, and every process contains both structural and agentive elements. This analytical approach regards social structures as the result of symbolic and instrumental human action, rather than as the fixed and given categories through which all social life is organized. The approach enables an approach to human social life that assumes constant flux, negotiation, and interpretation by human agents with economic, political, and cultural interests.

15. See, for example, the film *The Urban Explosion*, produced and distributed by the National Geographic Society as part of its Journey to Planet Earth film series. See http://www.screenscope.com/journey/journey_urban.html (retrieved February 3, 2008). Studies of the cities of the global North, in turn, tend to focus on

issues of overconsumption, greenhouse gas emissions, and other conditions associated with relative wealth. At both poles of the binary, our expectations and priorities are preconditioned by the categories "North" and "South."

16. See also Tanaka (1997) for an extensive demographic portrait of slum housing and residents in Kathmandu. Bal Kumar K.C. (1988) reports that the proportion of people living in urban centers went from 4 percent in 1961 to 9.2 percent in 1991 (or from 1 in 25 to 1 in 10); further noted is that less than 50 percent of the urban population have access to adequate drainage, solid waste facilities, or sewage and sanitation services. K.C. describes a trend of "gradual shifting of the rural poor to urban areas," particularly from the "immediate countryside" areas of Ramechhap, Trisuli, Dolakha, Sindupalchowk, and Dhading. In 1991, informal settlements in urban riverbank areas were estimated to be growing at 12 percent annually, a rate twice that of the city itself (HMG/ADB 1991). In 2001, the growth of squatter settlements continued at a rate of 12–13 percent (Hada 2001:154). Between 1990 and 2000, the number of total urban squatter settlements almost tripled, with a majority located on public lands along rivers (ibid.).

17. While a megacity is a city with over ten million inhabitants, a hypercity has a population of over twenty-five million.

18. We might usefully question the historical exceptionalism that frames many policy discussions of inadequate and informal housing. Claims to a new and unprecedented historical moment, in which the majority of the world's population lives in cities, may have the unintended effect of detaching diagnoses of the present from deeper historical conditions that lend traction to particular housing conditions in specific cities.

19. See, for example, WRI 1996, which introduced "the urban environment" as a comprehensive set of global problems for the policy and academic audience of the World Resources Institute. For more recent representations of the city as an ecosystem and as an environmental problem, see Alberti et al. 2003; Collins et al. 2000; Parlange 1998; Pickett 1997; Pickett et al. 2001. Also see the February 8, 2008, issue of *Science* magazine, "Reimaging Cities."

20. See, for example, Davis 2006. Throughout the work, Davis develops a definition of the category of the cities of the global South as those that knit together his truly global range of case studies and examples.

21. Studies of the cities of the global North, in turn, tend to focus on issues of overconsumption, greenhouse gas emissions, and other conditions associated with relative wealth. At both poles of the binary, our expectations and priorities are preconditioned by the categories North and South.

22. Dawson and Edwards (2004: 2) assert, "If a turn to a discourse of the global South is to offer a useful intervention, a new cartography (rather than simply a more palatable term for the 'third world'), then the term South must indicate a critique of the neoliberal economic elite and its management of the globe according to a developmentalist paradigm."

23. As Peter Hall (1988) writes in his classic history of planning, *Cities of Tomor-*

row, anxieties about the adverse social and environmental effects of urbanization animate urban planning history from its inception. Ebeneezer Howard's Garden City (1965 [1898]) is one example of how early urban planners considered natural spaces as remedies for urban social ills. Mike Davis (2006:5) writes that "Mumbai . . . is projected to attain a population of 33 million, although no one knows whether such gigantic concentrations of poverty are biologically or ecologically sustainable."

24. See, for example, "World Faces Population Explosion in Poor Countries," *Guardian Unlimited*, August 18, 2004, http://www.guardian.co.uk/population/Story/0,,1285358,00.html.

25. All forms of knowledge emerge within particular sets of social relations and institutional dynamics. Knowledge is not simply the revelation of facts; it is itself productive and reproductive of specific social and power relationships. Some useful elaborations of expert knowledge in political ecology and science and technology studies include Blaikie 1985; Bocking 2004; Brookfield 1999; Bryant 1998; Collingridge and Reeve 1986; Davis and Wagner 2003; Dimitrov 2003; Jasanoff and Martello 2004; Mitchell 2002; van Buuren and Edelenbos 2004.

26. Attempts to conceptualize how natural and social factors come together in cities include the Burch-Machlis model of the human ecosystem (Machlis, Burch, and Force 1997), which posits that human and nonhuman "hybrid characteristics" are found in most urban ecosystems. Unlike previous models that suggested reciprocal interconnections between a human and a natural ecosystem (Boyden 1993), this model uses more fluid nature–culture categories, suggesting that "some (ecosystem) components, fluxes, regulators, and processes in (urban) systems retain many of their 'natural' behaviors, whereas others may be entirely altered or constructed by humans" (Pickett 1997: 189). The Burch-Machlis model also explicitly recognizes "key hierarchies" in social organization (wealth, education, status, property, and power) and the difficulty of representing these at a variety of spatial scales. Thus, urban ecology has become the theoretical terrain in which longstanding disciplinary divisions between the natural and social sciences have begun, perhaps by necessity, to dissolve.

27. In this way, disciplinary urban ecology complicates the conventional Latourian critique of modern science and its "purification rituals."

28. See Swyngedouw (2004: 15–20) for an elaboration of the "production of space" as it might be applied to the environment.

29. A great deal of scholarship elaborates on this point; among it is work in science and technology studies that has paid productive attention to the social dynamics of scientific knowledge production (Downey and Dumit 1997; Dumit 2004; Franklin 1997; Hogle 1995; Stengers 1993; Rabinow 1992) and the ways that situated actions and contingent decisions compose scientific work (e.g., Knorr-Cetina 1981; Latour and Woolgar 1986). Other work in this vein has shown that technical problems are often defined in relationship with the spaces in which, and the processes through which, knowledge about the problem is produced (Callon

1995), and explored disciplined ways of organizing and making sense of the natural world (Barnes, Bloor, and Henry 1996; Gooding 1992; Lynch 1985).

30. See Alley 2002:26–29 for a discussion of whether culture or ecology deserves the dominant analytical role in matters of environmental studies.

31. Comaroff and Comaroff 1991:199; Latour 2005.

32. Following Riles's use of the "network" in *The Network Inside Out* (2000).

33. These works include but are not limited to Kramrisch (1976) for temples, Moore (1990) for Nyar houses in Kerala, Ifeka (1987) for Goan houses, MacFayden and Vogt (1977) and Gutschow and Kolver (1975) and Levy (1990) for the city of Bhaktapur in the Kathmandu Valley, Slusser (1982) for the Kathmandu Valley as a whole, Shepherd (1985) for Newar culture in general, and Gray (2006) for domestic space among the Kholagaun Chhetri of Basnaspati, Kathmandu.

34. Setha Low (1999:6) identifies two main bodies of urban anthropology research that foreground questions of ethnicity in the urban context. These are "1. The ethnic city as a mosaic of enclaves that are economically, linguistically, and socially self-contained as a strategy of political and economic survival; and 2. Studies of ethnic groups that may or may not function as enclaves but are defined by their location in the occupational structure, their position in the local immigrant social structure, their degree of marginality, or their historical and racial distinctiveness as the basis of discrimination and oppression."

35. Aspirations to a unified meaning are captured in such landmarks of the global environmental movement as the Bruntland Commission's *Our Common Future* (1987) and the 1992 Rio Declaration and Earth Charter.

36. See for example, Dixit and Ramachandran (2002); Gellner (2003, 2007, 2008); Gellner and Hachhethu (2008); Thapa (1994); Hutt (2004); Ogura (2008); Thapa and Sijpati (2004).

37. See for example Dixit and Ramachandran (2002); Gellner and Hachhethu (2008); Hachhethu (2002); Hoftun, Raeper, and Whelpton (1999).

38. However, scholarship on Newar identity stresses contestation and difference, and the many Newar activists and bureaucrats described in this book are unlikely to agree to an automatic and simplistic sameness based on their common Newar identity.

39. Gellner goes on: "No one—not even the most ardent Newar cultural nationalist—believes that they all derive historically from one place or one gene pool. They are one people, with a common language and culture, but they are divided into twenty or more castes, each of which has its own myth of origin" (2003:76).

1 CREATING NEPAL IN THE KATHMANDU VALLEY

1. Nepālmandala means "circle of the country of Nepal" (Slusser 1982). A mandala is a "ritual diagram with a principle deity at its center and the other divinities of this deity's retinue arranged geometrically around it. The mandala is the model for the design of Nepal's square pagoda temples and also, though less obviously,

for the layout of the royal cities of the Valley during the Malla period" (Hutt 1995:229).

2. As Bledsoe (2004) shows, the architectural history of the Kathmandu Valley is a relatively new area of study, having become a topic of interest and inquiry only in the mid-twentieth century. "Because there had been nothing akin to the institutions of colonial India—no Archeological Survey, no census—the field was wide open for international scholarship" (3).

3. See Hutt (1995:17–20). The Licchavis came from India and used Sanskrit as their court language. They built a strong centralized state mentioned in the history of northeast India. Although politically independent from India, they maintained southern ties; they also had political, economic, and cultural ties with Tibet. Licchavi rulers may have ruled a kingdom that extended beyond the confines of the Kathmandu Valley.

4. "Malla" does not refer to an ethnic or dynastic name but rather refers to "victors" and was assumed by the kings themselves. The early Malla period is not well documented, and little is known about how the first Malla king, Ari Malla (ruled 1200–1216) managed to supplant the Licchavis. By the fourteenth century there is more documentation. The Mallas became devotees of Taleju, the deity brought by the family of Harisingh Deva, the last Maithil king. Between 1200 and the mid-fourteenth century there were several raids and attacks on the Valley, but the weakness and vulnerability of the Malla kingdom appears to have ended around the reign of Sthiti Malla, whose rule set a century of relative strength and stability in motion. Sthiti Malla codified Hindu caste laws and oversaw the emergence of Newari as an important literary language and the addition of Degutale, a form of Taleju, taken in addition to Pashupati as the lineage's deity. Within the Gorkhali-Shah era is embedded the Rana era, which lasted from 1846 to 1951.

5. Bhaktapur is also known historically as Bhadgaon; Patan is also called Lalitpur. Later, the King of Nepal would derive his income primarily from the agricultural rents on the cultivation of crown land—assessed in cash (Burghart 1987:249).

6. Slusser (1982) writes, "In broad terms the Malla kingdoms were actually minuscule city-states; each kingdom consisted of its capital city . . . each capital city was actually a walled fortress . . . the three kingdoms were ruled by a bewildering number of kings . . . except for some of the last kings of Patan, rulers of the three kingdoms had common ancestry to Sthitimalla but their relations were essentially antagonistic and unpleasant." Prithvi Narayan Shah was honorably received by the court of Kathmandu in 1678 and he visited again in 1686. He embarked on the conquest of winning the Malla realms in 1685: that year Gorkha made the first of many alliances that would pit one Malla ruler against another. Meanwhile, Shah slowly conquered territory around the Valley perimeter (this began in 1744 with the seizure of Nawakot) and took command of the trade route to Tibet and northern access to the Valley. "During the next twenty years the little Valley was completely encircled by Gorkhali holdings, and underwent a debili-

tating economic blockade. Seeking relief, the ever-turbulent leaders of Patan entreated the Shah king himself to take its throne" (Slusser 1982:65). Finally in 1768 Prithvi Narayan Shah seized Kathmandu, then Patan, and, after more than a year, Bhaktapur—ending the Malla Era.

7. Kathmandu was not Prithvi Narayan Shah's sole seat of authority, however. In keeping with Gorkhali practices, the court moved seasonally, spending winters in the Kathmandu Valley and summers in the hills at Nuwakot. Burghart (1984) maintains that the two capitals were seen as located in two separate "countries" (*deś*), and transit between them allowed the king to avoid appearing to conflate them. Prithvi Narayan Shah is also known to have preferred the Hill country, having expressed plans to build a capital to replace Kathmandu just before his death (Burghart 1984). Movement between the two capitals continued until the Anglo-Gorkhali War (1814–16) made travel unsafe for the court. With the Treaty of Segauli, which set the terms for war settlement, the custom of maintaining seasonal capitals was discontinued, and the Nepal Valley became the fixed seat of Gorkhali authority. Non-Valley territories were thus made provinces (*pradeśa*) and were meant to contrast with a set urban political and administrative center (Burghart 1984).

8. Particularly *Pode, Cyāme, and Kasāin*: The three walled cities of the Valley are regarded as having followed a general pattern of settlement in which "high castes tended to cluster around (an) exalted nucleus (the large central area where the palace was), the lower castes lived progressively further away, and, outside the wall, were the outcastes. Finally, well beyond the city wall laid the realms of the dead, the *śmaśāna* (Nepali, *masān*), the various cremation grounds and ghats. Superimposed on such human orderings were various other orderings related to the divinities. These were in the nature of mystic diagrams, mandalas in which particular sets of deities were linked in concentric rings of protection inside and outside the city" (Slusser 1982:94).

9. As Burghart notes, Prithvi Narayan Shah "claimed that the Hindu rulers and nobles of the plains had given themselves up to the enjoyment of pleasure so that they no longer possessed the ability to preserve their independence from the British" (1984:115–16).

10. Liechty (1997) notes that the terms of reference for the British and Muslims were one and the same during this period; their designation as being impure created for them a single category.

11. The Anglo-Gorkhali War (1814–16) was fought over territory; General Bhimsen Thapa's encroachments into Company land provoked a formative response and the British easily defeated Thapa's army. The settlement outlined in the Treaty of Segauli (1816) reduced Nepal's territory and established the presence of a British Resident in Kathmandu. Thapa is said to have followed the conditions of the treaty meticulously but to have maintained as far as possible a policy of isolation and nonengagement with foreigners of any kind, including the British Resident. But by the outbreak of the Anglo-Gorkhali War, trade through Nepal

was no longer a major consideration for the British, as the Kathmandu route had become less economically important to the British than opium and cotton pursuits in China. That the importance of promoting trans-Himalayan trade had diminished likely affected the terms put forth by the British in the Treaty of Segauli. In the end, in addition to a Resident, the treaty demanded a clearly demarcated border between India and Nepal and an assurance that Nepal would not violate it. It stopped short of establishing clear political domination by the British, a condition that may have been included had perceptions of the economic importance of the Nepal Valley not changed so significantly.

12. Despite their isolationist policies, the Ranas themselves enjoyed consuming foreign luxury goods, well documented by Liechty (1997).

13. Political power was increasingly consolidated in the hands of regents and army commanders from about 1777 onward; this was due to the fact that several kings were minors or were incapacitated during this time. Two families, the Pandeys and the Thapas, feuded between themselves to attain and keep the prime ministership. A powerful army general and leader, Bhimsen Thapa, was overthrown in 1837, bringing about near chaos among those who would wish to rule. By 1846, Jung Bahadur Kanuwa brought the feuding to an end by murdering nearly all of his political competitors in the Kot Massacre. Jung Bahadur retained control until he died in 1877; during the course of his rule he set up a system of a hereditary prime ministership in which the Shah kings, although still on the throne, acted in practice as figureheads. Until 1945, the prime ministers of Nepal were all nephews of Jung Bahadur. A close alignment with the British characterizes the Rana era; the Ranas sent Nepali soldiers to aid the British in 1857 and allowed Gurkha soldier recruitment to operate inside Nepal from 1887 onward (it had been taking place outside Nepal since 1815). With this as the exception, isolationist policies were strictly enforced while most of the enormous tax revenues amassed by the government were put to personal rather than state use. An elaborate architecture, courtly culture, and military were features of Rana rule.

14. In the logic of the *Muluki Ain*, crimes and punishments were defined as much by a person's caste as by the act itself; for instance, in accordance with Muluki Ain a Brahman could not be put to death under any circumstances.

While a caste system was officially codified through the Muluki Ain, the idea of caste had been used by Shah rulers and elites as an overarching conceptual framework for ordering diversity in Nepal throughout the Shah era as well. Rajendra Pradhan (2002:7) notes that during the Shah era the primary concerns of the state were the exercise of political control and the extraction of revenue; creating a uniform national culture was not a priority. "At the same time," he writes, "the elites did need an overarching framework to integrate the diverse communities of the newly-expanded kingdom, and also to establish it as a pure and true Hindu land, an *asal* Hindustan, as Prithvi Narayan Shah called it" (7). This model was the caste system, which was used to organize and order Nepal's diversity. By the mid-nineteenth century the new ruling elites were "firmly ensconced in the palaces of the Malla kings in Kathmandu" (7). "The kingdom became more integrated and centralized, both politically and administratively. Meanwhile, with the blessing of the state, the land-hungry Parbatiya populace began migrating in ever-larger numbers to territories populated by ethnic communities. While the rulers were mainly concerned about extracting revenue for themselves and in consolidating their hold over the kingdom, as time went by they became increasingly interested in imposing a more homogenous cultural matrix on the kingdom as well" (8).

15. The hierarchy was arranged as follows: (1) high-caste Hindus (*tagadārī*); (2–3) alcohol drinkers (*matwālī*), divided into enslavable and unenslavable; (4) impure but touchable (*pānī na chālne*, cannot share water with higher groups); and (5) untouchable (*pānī na chālne*).

Categorization was often blunt and conflated diverse languages and cultural groups into broad caste-ethnic labels. The matwālī category collapsed a great range of people who had historically spoken Tibeto-Burman languages: groups seen as allies of the state, like Gurungs and Magars who had served in the army, were placed in the unenslavable matwālī category, while those marginal in the affairs of the state were assigned the enslavable ranking. Limbus were moved from enslavable to unenslavable in recognition of their services in the 1856 war with Tibet (Hofer 1979:124).

16. This points to the importance of Hinduization, or Parbatiyasition, as a cultural current within Nepali state making processes. Rajendra Pradhan (2002:5) notes,

> What is called the Hinduization of Nepal was actually the Parbatiyasition, that is, the spread and imposition of the culture of the Parbatiya, most significantly their language, Nepali (originally known as *Khas* or *Khas Kurā*), and religion, Hinduism. Even though Kathmandu Valley itself had long been a center for Hindu devotion and pilgrimage, it was the Gorkhali kings who spread the faith in its diverse forms across the mid-hills of Nepal. The process of Parbatiyasition was, and to a degree continues to be, facilitated by the state, because a majority of the ruling elite since the time of King Prithvi Narayan have been "high caste" Parbatiyas, who were actively supported by the Newari elite of Kathmandu, the majority of whom were also Hindu. The subordinated communities responded with accommodation and assimilation, but also with out-migration or resistance, sometimes violent.

Sudhindra Sharma (2002:25) identifies Hindu religious identity as an enduring feature of state assertions of religio-cultural purity, quoting Jung Bahadur Rana as saying, "In this age of Kali this is the only country where Hindus rule." Sharma writes, "Hindu religious identity was thus used as a means to assert Nepal's cultural uniqueness vis-à-vis Hindustan during the formative period of the nascent nation-state" (ibid.).

17. The boundary of the realm and the boundary of territorial possessions were aligned in 1860.

18. In November 1950, King Tribhuvan, who had been kept as a virtually powerless figurehead by the Rana prime minister, fled Kathmandu and sought political

asylum. The move catalyzed the demise of the Rana regime and by February 1951 King Tribhuvan returned as ruler. The political instability that followed delayed elections for a constituent assembly. Upon Tribhuvan's death in 1955, his son Mahendra became king, and Nepal established important international ties, including diplomatic links with the United States and contacts with China. Nepal also became a member state of the United Nations. By 1959, King Mahendra proclaimed that free elections would be held, which resulted in an overwhelming victory for the Congress Party. After an outbreak of violence in Gorkha, King Mahendra invoked emergency powers, dismissed the elected prime minister Bisheswor Prasad Koirala, and arrested elected members of Parliament and the cabinet. Mahendra announced that the democratic experiment had failed, and he replaced it with his idea of a democratic system more suited to the needs of Nepalis. His Panchayat constitution was introduced in December 1962. The central power figure in the Panchayat was the king.

19. The idea of nationalism as a project, and the promotion of Nepali language as a unifier of a single nation, is linked to activists and intellectuals working against the Rana regime from exile in India (Fisher 1998; Onta 1996). These activists began to consolidate Nepali by writing literary works in Nepal and creating national heroes by reviving literary figures like the poet Bhanubhakta Acharya (Onta 1996).

Pradhan (2002:11) notes that "the first National Planning Commission went so far as to argue in the early 1950s that, 'if the younger generation is taught to use Nepali as the basic language then other languages will gradually disappear, and greater national strength and unity will be the result.'"

Geertz (1973) brought the discussion of identity mobilization into the specific context of the nation-state, illustrating the susceptibility of new states to internal conflict as resources are partitioned and lines of power are redrawn. This competition tends to catalyze a new self-awareness among ethnic groups and a fervent sense of difference from other groups sharing the territory that has become the nation. This is reflected in contemporary terms in current contestation between the monarchal Panchayat legacy and the various *janājātī*, or noncaste ethnic groups, in Nepal (Hangen 2000). Additionally, the "unity in diversity" dilemma that is common to nation-building efforts around the world necessitates a hierarchical valuation of the ethnically diverse groups that constitute the generic ("national") whole. The example of Radio Nepal is instructive: in developing national music styles and representations of news and the nation, it invented and coordinated national identity and tradition, necessarily foregrounding particular aspects of Nepal's ethnic landscape, marginalizing others, and creating a generally ahistoric composite.

20. King Mahendra introduced a new Muluki Ain in 1963, from which the caste system was entirely absent (Pradhan 2002:12).

21. Joshi (1997) reports that between 1951 and 1997 the Nepali government received about US$3.7 billion in development funding.

22. See Rademacher and Tamang 1993:14–15.

2 KNOWING THE PROBLEM

1. In general, narratives play a vital role in discursively constituting moral agents and giving meaning to individual and collective actions. They also designate boundaries between nature and culture (e.g., Cronon 1995).

2. See Stanley International et al. 1994.

3. The prescription for restoration called for a widening of the river through weir dam construction, which would trap sediment and rebuild the river bed. Parkland development and temple rehabilitation along the riverbanks were considered crucial, as were better solid waste management, a sewage treatment scheme, and the removal or relocation of existing riparian sukumbasi settlements.

4. Other projects included the *City Diagnostic Report for City Development Strategy* (2001) by the Kathmandu Metropolitan City and supported by the World Bank; the *Nepal National Report* (HMG Nepal National Habitat Committee 2001), supported by the United Nations General Assembly on Human Settlements Istanbul +5 Meeting; the *Environmental Study of the Bagmati Watershed and Mitigation of River Pollution* (Nepal Environmental and Scientific Services 1997), which was published by the Ministry of Water Resources; *Urban Water Supply Reforms in the Kathmandu Valley: Wastewater Management Plan Assessment* (Metcalf and Eddy in association with CEMAT Consultants 2000); and the *Environmental Baseline of Bishnumati Corridor in the Kathmandu Valley* (Community Led Environmental Action Network [CLEAN-Nepal] 2002), published by World Vision International.

5. The Upper Bagmati Basin, a 662-square-kilometer area that includes the drainage of the Bagmati as far downstream as Bhandarikarka, beyond the Valley river outlet of Chobar, includes the Kathmandu Valley, which makes up 15 percent of the total basin area (Stanley International et al. 1994:1). The Middle Basin encompasses the remaining area of the Bagmati in the hill region, while the Lower Basin includes the Bagmati flow through the southern Tarai region, and into India, before it converges with the Ganges River. The Ganges, in turn, drains an area of 1,703,000 square kilometers, flowing about 2,700 kilometers through China, India, Bangladesh, Bhutan, and Nepal. The river reaches of greatest concern were those of the Upper Bagmati Basin, which flow through the capital city. A Kathmandu NGO, Environment and Public Health Organization (ENPHO), has monitored drinking water quality in Kathmandu since 1988, consistently reporting a chemically but not bacterially safe water supply (R. Shrestha n.d.: 5, 7).

6. The agency overseeing municipal water delivery at the time was the Nepal Water Supply Corporation. Surface and groundwater statistics come from R. Shrestha n.d. No license is required for private groundwater extraction, and data on private wells and their discharges have been estimated only in small-scale surveys and from site measurements. Despite a lack of comprehensive data, policy documents often state that groundwater extraction rates exceed recharge rates and are hence unsustainable. One estimate suggested that a high rate of private

extraction is causing the water table in the Valley to be lowered at an average rate of 2.5 meters per year (Stanley International et al. 1994, appendix 5:8).

7. Metcalf and Eddy in association with CEMAT Consultants (2000: II-2) cites a figure of 46 percent of drinking water as being supplied from groundwater sources.

8. Spatial and seasonal climatic variations affect river water levels and riverbed morphology. Between 70 and 80 percent of Kathmandu's rainfall occurs during the monsoon period, from mid-June to late August (Stanley International et al. 1994, appendix 5:3), but even during the monsoon, rainfall in the Upper Bagmati Basin is not uniform. Minimal river flow occurs in the dry months of April and May; while flow increases with the monsoon season and peaks in July and August. Metcalf and Eddy in association with CEMAT Consultants (2000: II-3) report that "the existing water supply situation in the Kathmandu Valley is critical: some households receive water for only 1–2 hours per day in the dry season, and some receive none at all. The water shortage is compounded as the population in the service area continues to grow. The water supply shortage is further magnified by this year's (2000) record dry season. Many commercial and industrial establishments and homes have installed tube wells to supplement or totally meet their water requirements." See also pages II-4 and II-5 of this report.

9. R. Shrestha (n.d.) estimates that Valley water demand is about 200 million liters per day (MLD), with an average per capita consumption of 107 liters per day (5). Shrestha also points out that municipal water production is only about 121 MLD in the rainy season, and about 88 MLD in dry months. This, combined with a piped water source leakage rate of up to 40 percent of the total production in the supply system, illustrates the need for supplementary groundwater use on a private basis.

10. See Stanley International et al. (1994:A7) for an inventory of studies and summary of their findings. Major studies include Shrestha and Sharma 1996, Pradhan 1998, and Metcalf and Eddy in association with CEMAT Consultants (2000).

11. pH is a value that indicates the acidity or alkalinity of a solution on a scale of 0 to 14, based on the proportion of H+ ions present. The DO is the amount of oxygen dissolved in a given volume of water at a given temperature and atmospheric pressure; it is usually expressed in parts per million. The BOD is a standard test for measuring the amount of dissolved oxygen utilized by aquatic microorganisms over a five-day period. The effects of oxygen-demanding wastes on rivers depend on the volume, flow, and temperature of river water. Aeration occurs in turbulent, rapidly flowing water, making it able to recover from oxygen-depleting processes. Concentrations of 100–150 mg/1 are typical of domestic sewage, so dry season flows highly resemble wastewater as the Bagmati passes through the urban core (Metcalf and Eddy in association with CEMAT Consultants 2000:II-1). Coliform bacteria live in the intestines of humans and other animals; the total coliform measure is used to determine levels of feces in water or soil.

12. Concentrations as high as 20 million FC/100 milliliters were recorded in the

Tekucha Khola in 1994; the same sampling cluster taken in the dry season found 9 million FC/100 milliliters in the downstream Bagmati and 5 million FC/100 milliliters at Pashupatinath (Stanley International et al. 1994, appendix 7:14).

13. The reduced BOD levels at Sankhamul (Station 5) are attributed to inputs from the Manohara, a tributary that converges with Bagmati just below Shankhamul. Manohara is relatively cleaner than the Bagmati. Dilution effects from the tributaries Balkhu, Nakhu, Bosam, and Chalti help explain reduced BOD at Chobar. See Shrestha (n.d.): 2.

14. From its origin in the Mahabarat Hills, the Bagmati cuts through phyllite, quartzite, and dolomite formations. In the Tarai, it crosses the main boundary thrust and Churia Hill formations. The Bagmati's upper, middle, and lower reaches have distinct characteristics relevant to sediment transport along its length. Upstream of Sundarijal represents a mountainous stage with a steep slope in ratio of 1:6.5. In this section, the riverbed is composed of gravel and boulders. Below Sundarijal, the river meanders and moves southward. Here, the slope begins to flatten to a ratio of about 1:285, which continues in the lower reaches until Khokana (approximately at the outlet of the Valley), after which the river slope increases as it flows through the Siwalik to 1:47. Once the river reaches the Tarai, the slope averages 1:540 (Stanley International et al. 1994, appendix 6:6).

15. Nepal Environmental and Scientific Services (1997: 3–7) reports: "There is very little information on sediment transport capacity of the Bagmati River and its tributaries. From the hydrological characteristics of the tributaries of the Lower Basin, it is assumed that these streams have the highest sediment transportation capacity compared to other streams of the watershed." Also, "Looking at the present morphological conditions of the rivers of the Kathmandu Valley, a complete ban on river bed sand mining should be imposed. Construction sand for the Valley can be mined from suitable degraded hillocks of the Valley without causing adverse impact on the surrounding environment" (1997: 5–3).

16. Carpet dyeing, for example, requires ammonium sulfate, urea, metal compounds, sulfuric acid, chromium, copper, iron, tin, hydroxides, and acetic acid; carpet washing uses chemicals including bleach, sulfuric acid, and ammonium sulfate. While there are no data on precise inputs of pollutants from the carpet industry specifically, it can be inferred that the industry plays an important role in river pollution in some reaches of the system (Stanley International et al. 1994, appendix 7:13).

17. Alley writes, "When Hindus and scientists use the word *pollution* or the Hindi transliteration *pradusan*, they are generally invoking as well as reinventing different conceptual constructions. These conceptual constructions are also, in some cases, different from the environmentalist's and the anthropologist's notions of pollution and impurity. Residents of the field sites I cover in this text revealed this conceptual pluralism when using the term pollution to talk about waste and the people who manage it. Transliterating the English term pollution in Hindi with *pradusan*, ... [they] ... associated a person rather than an ecological

process with the condition of pollution" (22). Alley's work goes on to elaborate and contrast complex conceptual elements of pollution as they were considered and used by variously positioned actors in the ethnography.

18. These include the classic works of Feldhaus (1995) and Gold and Gujar (2002), and a body of work that sought to outline a historical concern for nature derived from Hinduism (Chapple 1998; Chapple and Tucker 2000; Cremo and Goswami 1995; Narayanan 2001; Nelson 1998).

19. As Sudhindra Sharma (2002) has shown, there is a long history of interconnection between Nepali statecraft and religious symbolism. Sharma argues that at particular historical moments, the state has sought to enhance or consolidate its power through formal associations with religious symbolism and practice, citing for example the post-1990 context, in which, via the continued constitutional existence of Nepal as a "Hindu Kingdom," Hinduism functioned primarily as a force for the cultural legitimacy of the state. He cites examples like the ban on cow slaughter, official celebration of Hindu holidays, the content of school curricula, and the ban on religious proselytizing (30) as evidence of state-sponsored Hinduism. The mandala and other symbolic content of Maitighar may be interpreted as being consistent with this.

20. According to the World Resources Institute (1996), the urban growth rate in the cities of the Kathmandu Valley was 7.1 percent over the period 1990–95, a figure considerably higher than the United Nations Population Fund (UNFPA) estimate. The WRI study estimates that by 2025 the percentage of Nepal's population residing in urban areas will increase to 34 percent, from the present 14 percent. K.C. (1998) reports that the population of Nepalis living in urban centers went from 4 percent in 1961 (or 1 in 25 people in cities) to 9.2 percent in 1991 (or nearly 1 in 10 people in cities).

21. Even by the early 1990s land values in the city core were skyrocketing: the HMG/ADB report in 1992 estimated that the land in the Bishnumati Corridor was worth at least Nepali Rs 50,000 (U.S. $800) per ropani or Rs 98 million per hectare (U.S. $1.6 million). This report also estimated that these values were increasing by 20 percent annually (33).

22. Indeed, the very notion of the "rural in the urban" is only recently attributable to the presence of rural-to-urban migrants in the city. Previously, such a discussion might involve the "rural" settlements inside the Kathmandu Valley that were gradually "swallowed up" by urban expansion. See, for example, Anup Pahari, "The Villagers of the Valley," *Himal* 5(1):13–16.

23. In the fall of 1997, the time of my master's-level research, the total number of settlements characterized as sukumbasi in Kathmandu was fifty-four. Half of these were riparian and were situated on the banks of the Bishnumati, Bagmati, or one of their larger urban tributaries (Tanaka 1997). Of the total population of sukumbasi in the Kathmandu Valley in 1996—close to 9,000 (ibid.)—69 percent lived in riparian zones and about two-thirds of those residents occupied settlements on the banks of the Bishnumati or Bagmati rivers. By 2001, Hunt's pub-

lished data (2001) showed an increase in the number of urban squatter settlements to sixty-five.

24. Contrary to popular generalizations about the caste composition and place of origin of riparian sukumbasi, extensive household-level data collected by Tanaka (1997) demonstrated vast caste and ethnic diversity in the settlements. A later study, by Hunt (2001), reinforced Tanaka's findings and demonstrated that the primary motivation for migration among sukumbasi was a lack of sufficient land or employment in their place of origin. Hunt's data differentiate between "slum" communities, made up almost exclusively of Dalits, and "squatter" communities, made up largely of non-Dalits who have migrated to the Valley. Hunt argues that historically, the "binding element in the squatter communities has not been caste, but rather that of newcomers needing to find and build a new community" (Hada 2001:147). There are a few exceptions, including low-caste settlements along the Bishnumati: "Ever since the Malla period, diverse castes and ethnic groups, primarily Newars, were concentrated in various areas of the city, either in specific sectors or in radial clusters centered from the city core. Lower castes and untouchables were on the edge of the communities, particularly toward the two rivers. The communities of these scheduled castes still exist along the Bishnumati River corridor" (Hada 2001:149). Hada found a high representation of Hill Ethnic groups and Bahun-Chhetris in squatter communities; in data collected in 84 percent of all squatter settlements, she reported an ethnic composition of Newar, 13 percent; Bahun-Chhetri, 27.7 percent; Hill Ethnic, 48 percent; Other, 11.3 percent. Of Hunt's respondents, 71.2 percent originated in the Western Development Region, and 14 percent were from the Eastern Development Region.

25. The Municipality Act of 1995 made Kathmandu a municipality—divided into thirty-five wards of which the city core has twelve wards. Much of the municipality is bounded by the Bagmati River and Ring Road; its total area is 5,076 hectares with an estimated population of 729,690 and 5.65 percent growth in 2001 (Hada 2001:150).

26. See Surendra Phuyal, "Welcome to the World of Squalor Amidst Plenty Right Under Your Nose," *Kathmandu Post*, 1999.

27. The Healthy Cities Program, launched by the World Health Organization (WHO) European Regional Office in 1986, used a public health approach to physical and social environments. It emphasized health promotion as a way of enabling, facilitating, and mediating development and environmental change. The Sustainable Cities Programme (SCP) is a joint initiative of the United Nations Centre for Human Settlements (HABITAT) and the United Nations Environment Programme (UNEP). Established in 1990, it aims to improve the environmental management and planning capacity of municipal authorities and their affiliates. The term "sustainable cities" has also been used by other international agencies, including regional development banks and the World Bank.

28. A "Healthy City," for instance, as it was explained at the World Habitat Day Proceedings in Kathmandu, requires a "healthy environment," assessed in

its capacity to provide food, clothing, and shelter (Bajracharya and Manandhar 1997:5).

29. I sought to investigate this assertion in my own research through a formal survey conducted in four riparian sukumbasi communities in 1997 with the assistance of Jeevan Raj Sharma and Dorje Gurung. Focus group discussions were also conducted in the settlements on the topics of river history, condition, and restoration. See Rademacher 1998: 46–63.

30. Sukumbasi advocates did not function independently of the international development community; in fact, Lumanti depended in large part on donor funding. This overlap highlights a point made at the outset: analytical lines drawn between contrasting framings of river ecology were often dynamic. While their ecological perspective may be categorized as "unofficial," Lumanti in particular had a definite link to a particular discursive and material current within the international development industry that emphasized human rights and managing urban growth.

31. While housing advocates acknowledged that resettlement could be an option, they doubted its actual viability on several grounds, including the large numbers of people who would have to be moved and the unwillingness of most rural-to-urban migrants to be relocated.

32. Huta Ram Baidya grew up in Thapatali, just across the road from the Bagmati River. Nepal's first agricultural engineer, Baidya served in the government sector for thirteen years. After leaving the government, he concentrated on several advocacy initiatives, including the Save the Bagmati Campaign, which he founded in 1990.

33. For a more detailed discussion of the Bagmati Civilization idea, see Rademacher 2007 and Rademacher 2009.

34. Driver (1988) suggests the phrase "ideal moral geography" as a way to analytically relate urban environments to the social production of people out of place, or outsiders. The moral component marks naturalized rules or expectations that are, in actuality, contingent and constructed; it underlines the degree to which moral geographies are also ideologies of belonging. Considerable scholarly attention has also been paid to the "moral geography" of Nepal and the Kathmandu Valley through elaborations of the structure and function of sacred landscapes, and this is exemplified through such works by Slusser 1982, Levy 1990, Gutschow and Kolver 1975, and Gutschow, Michaels, Ramble, and Steinkellner 2003.

35. The development industry in Nepal was highly influential throughout the Panchayat era, and it enjoyed continued importance during the contemporary democratic period (1990–2002). See, for example, Acharya 1992, Dixit 1997, and Seddon 1987.

36. *Images of a Century* contains a collection of photographs from 1919 to 1992, originally compiled by the Nepal Heritage Society and a GTZ-supported NGO called Urban Development through Local Efforts.

37. Jung Bahadur Kunwar, later Jung Bahadur Rana, was the first Rana ruler to rise to power following a massacre of all of his likely competitors in 1846.

3 EMERGENCY AND AN UNSETTLED CITY

1. Barbara Crossette, *New York Times*, June 2, 2001, 1.

2. Among the fatally wounded were King Birendra, fifty-five years old, and Queen Aiswarya, fifty-one; their son, Prince Nirajan, twenty-two; their daughter, Princess Shruti, twenty-four; the king's sisters, Princess Shanti Singh and Princess Sharada Shah; Princess Sharada's husband, Kumar Khadga Bikram Shah; and Princess Jayanti Shah, one of the king's cousins.

3. In later explanations for the suggestion that the incident was a single accident, government officials cited constitutional limitations on associating Nepali royalty with criminal activity.

4. Politically, Gyanendra was known to be, in many ways, quite different from his brother. A political hardliner, he was said to have adamantly opposed the democratic reforms implemented by Birendra in the early 1990s. Perhaps even more disturbing to the Nepali public was Gyanendra's now-forty-year-old son, Paras, whose notorious behavior was widely publicized. Just a year before, Paras was accused of vehicular homicide involving a popular Nepali singer. The public response to that incident was an unprecedented petition, said to have contained half a million signatures, appealing to the king to revoke Paras's royal privileges and try him as a criminal. At the time of his murder, King Birendra had not responded to the petition.

5. In the absence of information from the government, other rumored explanations of the murders circulated. One held that Prince Dipendra had secretly married his girlfriend, Devyani Rana, without the permission of his parents, who were warned by an astrologer that Prince Dipendra should not wed until he reached age thirty-five or his father would die. At the time, the prince was only thirty, so the king had to die. The other explanation conjectured that, long ago, Prithvi Narayan Shah offended an ascetic by vomiting when he was given yogurt. Instead of eating the yogurt again, Prithvi Narayan threw it away. Had he finished the yogurt, the Shah reign would have lasted forever, but in throwing it away, the reign was cursed to end after ten generations. Prince Dipendra was within the eleventh generation of Shah rulers.

6. The CPN-M's demands were reprinted in an English translation in the *Nepali Times* on February 15, 2001.

7. The three-member panel was to include Chief Justice Keshav Prasad Upadyaya, Lower House Speaker Taranath Ranabhat, and the leader of the main opposition party, the Communist Party Nepal-United Marxist-Leninists (CPN-UML). The inclusion of the latter, Madav Kumar Nepal, was initially seen as a gesture toward opening a legitimate discussion of what had happened in the palace, but such optimism was short-lived. Nepal withdrew from the commission on the following day, citing his opposition to the procedure that was followed to establish it. See Utpal Raj Misra, "Probe Body Suffers Setback as UML Leader Pulls Out: Move Questionable, Hints DPM," *Kathmandu Post*, June 5, 2001. Commen-

tators speculated that he had been pressured by other left parties, or perhaps his own, to withdraw.

8. Baburam Bhatterai, *Kāntipur*, June 6, 2001.

9. Yubaraj Ghimire, the *Kantipur* editor; Kailash Shiroiya, the managing director; and Binod Gyawali, a director, were arrested on June 6, 2001, in connection with the publication of an article by the Maoist leader Baburam Bhattarai. See "Nepal Journalists Charged with Treason," BBC South Asia News, June 6, 2001.

10. For discussions of the cultural idea of the Nepali kingship, see Bhatt 2002; Burghart 1987; Sharma 2002.

11. See John Burns, "Nepal's Royal Deaths Give Life to Swirl of Theories," *New York Times*, July 19, 2001.

12. See http://www.nepalnews.com (retrieved November 26, 2001).

13. See Onta 2002.

14. This particular sense of a "new" Nepal would likely take on a different form among rural Nepalis, or urbanites positioned in a lower-class or caste situation than my host family and middle-class, professional informants. Although I was not positioned in a way that allowed me to gauge these contrasts, it is reasonable to assume the existence of a large group of Nepalis for whom the "new" Nepal represented the possibility of political and cultural liberation or the shedding of historically entrenched symbols of oppression. This points again to the situated positions of my informants and my own perspective on the "crisis" of the emergency. Note that this claim to a new Nepal comes to its ultimate fruition only in the wake of jana andolan II. For a discussion of political transformation in the "new Nepal," see Hangen 2007.

15. Over the democratization period that began in 1990, newspapers emerged in a variety of languages, espousing a variety of political perspectives. This, combined with a proliferation of commercial radio stations and programs, created a popular media in the Kathmandu Valley. In particular, papers such as the Maoist-supportive *Janādeśa*, could be acquired and reviewed alongside publications with long histories as state organs (such as the *Gorkhapatra* and *Rising Nepal*) and those perceived to be more independent but less likely to directly challenge the existing political order (such as the fairly new *Nepali Times* or the *Kāntipur* and *Kathmandu Post* papers). *Janādeśa* was one of the first papers to be shut down during the emergency. See Onta 2002.

16. See, for example, "Clear Flows the Bagmati: One Step at a Time, the Valley's Major Source of Water Is Being Cleaned Up," *Nepali Times*, March 29, 2002; "A Sewer Runs through It," *Nepali Times*, April 20, 2001.

17. The committee formed in 1995.

18. The sewerage system covers an area of about 537 hectares and serves a population of almost 200,000. The plant sits on 95 *ropani* (a ropani is about 500 square meters) of land.

19. Another "master plan" to improve the Pashupati area specifically was underway at this time, administered by the Pashupati Area Development Trust (PADT).

This plan covers 264 hectares of land, divided into three sectors; the plan was organized in three phases: The first was related to clearing and acquiring land in order to create a protected area. The second phase involved resettling displaced families, preserving archaeological monuments, and building park-related infrastructure. The third phase of the plan involved soliciting international development assistance. This work was announced publicly in the fall of 2000; see "Work Begins to Clean Up Pashupati Area," *Kathmandu Post*, September 17, 2000.

20. A great deal of controversy preceded public euphoria over the improved water quality at Pashupatinath. In particular, the diversion tunnel was considered completely unnecessary by those who argued that the siting of the sewage treatment plant upstream of the temple meant that treated water would flow past the temple and do the same job the diversion tunnel was doing. Binan Tuladhar was quoted in the *Post* as saying, "The fact is, it seems treated water from the sewage plant is just not good enough for Pashupati. It's the same old story, crores and crores going down the drain to satisfy religious fanatics." The tunnel's price tag, estimated at Rs 89.3 million in 2000, was indeed a huge investment, particularly compared to the estimated cost of the plant—about Rs 50 million. Tuladhar also argued that the tunnel caused a delay in the completion of the sewage treatment plant itself (see Suman Subba, "Bagmati Waits Even as Government Ignores Clean-Up Call," *Kathmandu Post*, June 10, 2000).

21. The stated main objective of the project was to enact conservation initiatives on the Bagmati River from the Shivapuri watershed area to Chobar—in other words, to treat the Bagmati throughout its reach in the Kathmandu Valley. Its work was divided into four phases, the first and second of which involved the construction of drainage and sewers from Gokarna to Tamraganga and Mitrapark to Tamraganga; and the creation of a tunnel from Tamraganga to the Tilganga sewage treatment plant at Guheswari; and road and green belt areas from Gokarna to Tamraganga along the Bagmati River. Phases three and four were to work according to a master plan for the entire Shivapuri-to-Chobar river reach.

22. "Bāgmatī Dal Prashedakendramā Bishatra Padarya," *Kāntipur*, February 8, 2002.

4 EMERGENCY ECOLOGY AND THE ORDER OF RENEWAL

1. The SAARC summit was the first to be held in three years, and the second in Kathmandu since its creation in 1985. Far more than an alliance to promote regional trade relations, SAARC was conceived as a potential catalyst for the creation of "cooperative cultural identities" through which the political conflicts and tensions pervasive in the South Asian region could be "moderated, if not completely eliminated" (see Muni 2002).

2. This chapter addresses the dynamic ways in which urban nature is produced, following Cederlof and Sivaramakrishnan's suggestion that ecological nationalism "links cultural and political aspirations with programs of . . . environmental

protection, while noting their expression in, and through, a rhetoric of rights" (2006:6).

3. Many of the slogans were written in such a way that they were easily subject to satire, prompting Kunda Dixit to poke endless fun (e.g., Dixit and Ramachandran 2002).

4. Michael Hutt (1995:229) defines the mandala as "a ritual diagram with a principal deity at its center and the other divinities of this deity's retinue arranged around it. The mandala is the model for the design of Nepal's square pagoda temples and also, though less obviously, for the layout of the royal cities of the Valley during the Malla period." Hutt defines the stupa as "the primary cult object of Buddhism in Nepal and elsewhere." It was "originally a mound entombing sacred relics, now greatly elaborated" (1995:232).

5. Emergency restrictions did not legally prohibit affected individuals from filing court complaints, yet no complaints in the Thapathali eviction case were filed.

6. Several months after the SAARC conference, the *Nepali Times* newspaper ran a profile of Keshab Stapit, the Kathmandu mayor, that cited the transformation of Tinkune as evidence that the mayor lived up to his reputation as a "shrewd operator." The article explained, "When he had to convince Prime Minister Sher Bahadur Deuba to clean up Tinkune before SAARC, he told him that demolishing houses in the city would take media attention away from the Maoist insurgency ahead of the summit. Deuba immediately saw the logic, and gave the go-ahead."

7. See Sribhakta Khanal, "Tinkune and Maitighar," *Nepali Times* 73 (2002):7.

8. As Sudhindra Sharma (2002) has shown, there is a long history of interconnection between Nepali statecraft and religious symbolism. Sharma argues that at particular historical moments, the state has sought to enhance or consolidate its power through formal associations with religious symbolism and practice, citing for example the post-1990 context, in which, via the continued constitutional existence of Nepal as a "Hindu Kingdom," Hinduism functioned primarily as a force for the cultural legitimacy of the state. He cites examples like the ban on cow slaughter, official celebration of Hindu holidays, the content of school curricula, and the ban on religious proselytizing (30) as evidence of state-sponsored "Hinduism." The mandala and other symbolic content of Maitighar may be interpreted as being consistent with this.

9. See Baviskar 2003; Chakrabarty 2002.

10. See Lumanti Support Group for Shelter 2002a.

11. Among the official gestures that contributed to a growing sense of security among sukumbasi were municipality-issued identity cards distributed to families living in the settlements and visits from officials as prominent as Mayor Stapit only a few months before. The mayor made explicit assurances against evictions.

12. See Maarten Post, "Kathmandu's Malignant Urban Tumour," *Nepali Times*, September 6–11, 2003.

13. Before the emergency, rural-to-urban migrants were identified mainly as

economic refugees. In the context of the People's War they were increasingly framed as war refugees—or even, perhaps, as Maoist sympathizers and therefore political invaders. If prior to the emergency the riparian sukumbasi community was seen as an obstacle to reclaiming previous river ecology, during the emergency it was also seen as an obstacle to the very reproduction of the state, both in terms of security and in terms of the reproduction of material relations, such as by securing donor funding. In both cases, the cultural value and meaning of the rivers, interwoven with shows of state strength, were critical to definitions, and realizations, of ecological improvement. The national state of emergency declared in November 2001 reconfigured the political context within which debates over the urban environment took place. Previous narratives of ecocultural degradation and riverscape invasion now assumed overtly political inflections with important ties to the People's War. The "invasion" from the countryside was increasingly inflected with a particular tension. On one hand, more and more migrants were acknowledged to be refugees fleeing the brutality of the countryside; on the other, riparian settlements of migrants represented a relatively uncontrolled space where rebels might easily and secretly infiltrate the city. Thus even as we read or heard individual stories of desperate flight, the settlements themselves were cast in official and popular discourse as the potential refuges of possible "terrorists."

14. The UN Park, mentioned in the introduction, was first proposed in July 1995 by a group of acting and former state ministers, influential scholars, and development professionals to commemorate the fiftieth anniversary of the United Nations and Nepal's contributions as a member state. In its earliest formulation, the UN Park plan called for greenspace to replace sandy flats of the exposed riverbed on the banks of the Bagmati and Bishnumati. Part of the park's stated purpose was "to mark respect for those Nepalese army, police, and civilians who have died in UN peacekeeping endeavors." In its early stages, park boundaries fluctuated, but by 1999 a master plan covered river corridors along much of the Bagmati and Bishnumati in Kathmandu (Cosmos Engineering Services 1999:xi). From its inception in 1995 until 2004, however, the physical manifestation of the UN Park was largely unrealized.

15. Interview transcript, January 8, 2002.

16. Ibid.

17. Examples of these include Friends of the Bagmati (est. 2000), Save the Bagmati Campaign (est. 1990), and Nepal River Conservation Trust (est. 1995).

18. Regarding the Bagmati Civilization, see, for example, a recent comprehensive collection of Baidya's work: *Bagmati Sabhyata Samrakshana Sangharshaka Barha Varsha* (Kathmandu: Nepal Water Conservation Foundation, 2002); *Bāgmatī Sabhyatāko Garimā* (The Importance of the Bagmati Civilization), *Kāntipur*, December 2, 2002; *He Bāgmatī, Timi āū!* (Oh Bagmati, Please Come Back!) *Kāntipur*, 22 Jeth 2058 (June 4, 2001); and "The Endangered Bagmati Civilization of Kathmandu Valley" (posted online at http://www.users.cts.com/sd/a/abaidya/bagmati/histbagmati.com; retrieved in 1998; link no longer active). On May 16, 2000, Chaudhary's appeal ran in several Nepali newspapers.

19. *Kathmandu Post*, February 7, 2000.

20. Editorial, *Kathmandu Post*, February 8, 2000.

21. See also, for example, "Matter of Urgency" by Diwakar Poudyal, in a letter to the *Kathmandu Post* on April 23, 2000, and "Bagmati Needs Help, Not More Pollution," by Mukunda Thapaliya in a letter to the *Kathmandu Post*, July 8, 2000.

22. Note the stone-throwing incident and other instances of public humiliation of the king. In February 2007, a crowd of people in Kathmandu stoned the motorcade of the king, who escaped from the attack, unscathed. The incident was the first violent reaction to the king's rule since the protests in 2006 that caused him to renounce the self-proclaimed absolute power.

23. In some uses, the role of the monarch was differentiated through a democracy in which the king would be retained as a figurehead (loktantra), and a state with no monarchy at all (*ganatantra*). Susan Hangen noticed the use of loktantra among her informants in the ethnic political party, the MNO, in the late 1990s; she understood this usage as a way of referencing a vision of a more authentic democracy without a king (personal communication). By 2006, the term had entered prominent usage in the capital city.

24. In this period, the idea of a "new Nepal" came into prominent usage in the Kathmandu Valley as a way to differentiate the era and process of restructuring the democratic state from an "old Nepal" of emergency and, prior to this, prajatantra.

25. See Ramesh Raj Kunwar, "Bonded Labour: The Scenario in the Last Decade of the Last Millennium," *Voice of History* 15(1):35–64, in particular, the postscript to the piece (58–64). See also Guneratne 2002; INSEC 1992; Lowe 2001; Rankin 1999.

26. *Kathmandu Post*, July 19, 2006. Himalyan News Service, July 20, 2006, says that out of thirty thousand freed kamaiya families, only twelve thousand had been provided with land, and that land was in "inappropriate places." The families also were harassed and threatened with displacement.

27. Meanwhile, while protest in the capital was going on, kamaiyas in Banke, Bardiya, Kailali, Kanchanpur, and Dang districts placed padlocks on land reform offices there.

28. *Kathmandu Post*, July 27, 2006.

5 ECOLOGIES OF INVASION

1. My reference to this classic analytic follows Herzfeld (2006), who argues that Douglas's (1966) approach, that is, an analysis that attends to the symbolic meaning of those aspects of the social and material order deemed to be dirty or otherwise impure, is most useful when it incorporates "the dynamic aspects of spatial symbolism and the agency operating in and against bureaucratic power" (131).

2. Creswell (2005:128) calls a moral geography "the idea that certain people, things, and practices belong in certain places, spaces, and landscapes, and not in others." Anthropological literature on Nepal and the Kathmandu Valley includes

several important works on sacred geography, including, Gutschow 1985; Gutschow et al. 2003; Levy 1990; and Slusser 1982.

3. I use the phrase "informal housing" in the somewhat blunt way that it was used by activists and development professionals with whom I conducted participant observation; that is, it is generally applied to housing that is either untitled, or considered substandard, or both.

4. This point was made particularly vehemently in an interview with Prafulla Man Pradhan, of Urban Development through Local Efforts, 1997.

5. The Bishnumati Link Road Project was one element in a larger scheme to improve the Bishnumati Corridor in Kathmandu. It proposed a 2.8-kilometer road along the Bishnumati River that would link Kalimati and Sorahkhutte, with an aim to improve access to high-density areas and improve traffic flow. The concept dates to the Kathmandu Valley Physical Plan of 1969. In the 1970s a more detailed road plan was developed; reports in the early 1990s by the Asian Development Bank (ADB) and a Japanese investment group proposed routes for the road. In 1992, the ADB started three projects related to the Bishnumati Corridor, but in 1999 and 2000 ADB pulled out and the government assumed responsibility for the project. According to ADB, its decision was based, in part, on the government's refusal to compensate and resettle squatters who would be evicted for road construction.

6. Interview with Lajana Manadhar, January 8, 2002.

7. Hunt's (2001) data differentiate between "slum" communities, made up almost exclusively of Dalits, and "squatter" communities, made up largely of non-Dalits who have migrated to the Valley. Hunt argues that, historically, the "binding element in the squatter communities has not been caste, but rather that of newcomers needing to find and build a new community." There are a few exceptions, including low-caste settlements along the Bishnumati: "Ever since the Malla period, diverse castes and ethnic groups, primarily Newars, were concentrated in various areas of the city, either in specific sectors or in radial clusters centered from the city core. Lower castes and untouchables were on the edge of the communities, particularly toward the two rivers. The communities of these scheduled castes still exist along the Bishnumati River corridor" (Hada 2001:149). Hada found a high representation of Hill Ethnic groups and Bahun-Chhetris in squatter communities; in data collected in 84 percent of all squatter settlements, she reported an ethnic composition of Newar, 13 percent; Bahun-Chhetri, 27.7 percent; Hill Ethnic, 48 percent; and other, 11.3 percent. Of Hunt's respondents, 71.2 percent originated in the Western Development Region, and 14 percent came from the Eastern Development Region.

8. For example, after a March 29, 2002, explosion at the Bishnumati Bridge at Kalimati injured twenty-four people, all surrounding sukumbasi settlements were reportedly raided. In a January 2002 interview, the director of Lumanti reported similar raids in Balaju and Kumaristan after an explosion in the Balaju neighborhood.

9. Five communities were affected: Dhumakhel, ward 15, and Khushibahil, Cha-

gal, Tankeshor, Dhaukhel, ward 13. A total of 142 houses were scheduled for demolition.

10. According to the Lumanti Support Group for Shelter (2006:12), "People resettled anywhere they could find shelter, mostly in nearby areas. Many doubled up with relatives. Those who had been deemed eligible waited to receive their first rent payments, which did not materialize in many cases. Eligible residents in Dhumakhel received the money for three months rent. Those from Tankeshwor and Khushibahil were asked to provide assessments of their properties that had been made a few years earlier by the ADB, and those that could come up with them were provided with two months' rent."

11. Text of an e-mail message from a Lumanti volunteer, February 2002.

12. See Lumanti Support Group for Shelter 2006.

13. Ibid.

14. The three historical walled cities of the Kathmandu Valley are regarded as having followed a general pattern of settlement in which "high castes tended to cluster around (an) exalted nucleus (the large central area where the palace was), the lower castes lived progressively further away, and, outside the wall, were the outcastes. Finally, well beyond the city wall laid the realms of the dead, the śmaśāna (Nepali, masān), the various cremation grounds and ghats. Superimposed on such human orderings were various other orderings related to the divinities. These were in the nature of mystic diagrams, mandalas in which particular sets of deities were linked in concentric rings of protection inside and outside the city" (Slusser 1982:94).

15. I thank Arjun Guneratne for raising this point.

6 LOCAL RIVERS, GLOBAL REACHES

1. Hannerz (1996) suggests an analytical approach to the idea of globality as a force that creates spaces for interaction among things that were once disparate.

2. Historically, "local" has tended to refer to smaller scales of social organization, like communities or regions, while "global" has tended to imply both commonalities in human experience and social conditions shared over large scales. One problem with accepting these terms as given is the accompanying assumption that they are static; another is that "the local" implies a bounded place or social experience while the global seems unplaceable and unboundable. This has particular implications for the definition of problems, because it allows global institutions and flows to appear outside of, or removed from, situations understood to be "local." Bounding the local scale can obscure the ways that global institutions or processes are connected to local problems. Similarly, a conventional sense of the "global" is often derived, stabilized, and naturalized by specific institutions, and the universalizing narratives, ideologies, and critiques that circulate through, and emanate from, them (e.g., Peters 1996, Forbes 1999).

3. Cosmos Engineering Services 1999:xi.

4. The study area extended about fifty meters on either side of the rivers in each corridor. Cosmos Engineering Services 1999:xi.

5. Through its government affiliation with the Ministry of Housing and Physical Planning, the UN Park Development Committee, also called the Bagmati-Bishnumati Conservation Program, has managed to obtain some funding from the government—enough to keep the project afloat but never enough to implement more than the most basic elements of its overall management plan.

6. See, for example, Sailash Gongol, "Un-parking the UN Park," *Kathmandu Post*, September 29, 1996, which predicts, among other things, that the cemented edges that featured in the park plan would cause a "bottleneck" along the rivers and lead to flooding. See also "Disaster Unpreparedness" in *Nepali Times* (July 26–August 1, 2002), in which the UN Park area flooding is reported.

7. Field notes, interview with Laxman Shrestha, December 5, 2001.

8. Field notes, interview with Laxman Shrestha, January 8, 2002.

9. Bhim Sen Thapa was Nepal's third prime minister. In 1814 he declared war against British territorial aggression in the region. Thapa and his forces lost the war, which ended with the Treaty of Segauli in 1816. With this treaty, Kumaon, Garhwal, and Sikkim territories were lost to the British, along with a portion of the Terai to the south. A permanent British residency was established in Nepal, and British recruitment of Nepali soldiers to the Gurkha battalions was regularized. Following the war, Thapa returned to his duties as prime minister, and he set out to Europeanize the army, reform the civil administration, delineate district borders, assess land taxes, and perform a range of other state making tasks.

10. Wright 2000 [1877]:15, 264.

11. Huta Ram Baidya, "Suggestion," *Kathmandu Post*, January 29, 1996.

12. See http://www.arcworld.org/projects.asp?projectID=179 for an official description of the Bagmati as part of the Sacred Gifts for the Planet campaign.

13. See http://www.friendsofthebagmati.org.np.

14. Relative to most of Kathmandu, those Nepalis who participated in Friends of Bagmati meetings at this time were generally from privileged backgrounds; they were the city's elites in some combination of social, political, economic, or educational terms.

15. *Nepali Times* #56, August 17–23, 2001.

16. This day, December 29, corresponded with International Biodiversity Day and the king's birthday.

17. Field notes, January 14, 2002.

18. I knew that the women spoke and understood English, as they both spoke with me in English on prior occasions.

References

Abrams, P. 1982. *Historical Sociology*. Ithaca: Cornell University Press.

———. 1988. "Notes on the Difficulty of Studying the State." *Journal of Historical Sociology* 1:58.

Abu-Lughod, J. 1999. *New York, Chicago, Los Angeles: America's Global Cities*. Minneapolis: University of Minnesota Press.

Acharya, M. 1992. "In Aid in Nepal's Development: How Necessary?" Position paper from a panel debate organized by Udaya-Himalaya Network, Kathmandu.

Agrawal, A. 1996. "Poststructuralist Approaches to Development: Some Critical Reflections." *Peace and Change* 21(4): 464–77.

———. 2001. "State Formation in Community Spaces? Decentralisation of Control over Forests in the Kumaon Himalaya, India." *Journal of Asian Studies* 60:9–40.

———. 2005. *Environmentality: Technologies of Government and the Makings of Subjects*. New Haven: Yale University Press.

Alberti, M., J. Marzluff, E. Shulenberger, G. Bradley, C. Ryan, and C. Zumbrunnen. 2003. "Integrating Humans into Ecology: Opportunities and Challenges for Studying Urban Ecosystems." *Bioscience* 53:1169–70.

Allen, T., and T. Hoekstra. 1992. *Toward a Unified Ecology: Complexity in Ecological Systems*. New York: Columbia University Press.

Alley, K. 2002. *On the Banks of Ganga: When Wastewater Meets a Sacred River*. Ann Arbor: University of Michigan Press.

AlSayyad, N. 2004. "Urban Informality as a 'New' Way of Life." In *Urban Informality: Transnational Perspectives from the Middle East, Latin America, and South Asia*, ed. A. Roy and N. AlSayyad, 7–32. Lanham, Md.: Lexington Books.

Amatya, S. 1994. *The Bagmati: Between Teku and Thapathali: A Monument Guide*. Kathmandu: Modern Printing Press.

Amin, A., and N. Thrift. 2002. *Cities: Reimagining the Urban*. Cambridge: Polity.

Appadurai, A. 1992. "Putting Hierarchy in Its Place." In *Rereading Cultural Anthropology*, ed. G. Marcus. Durham: Duke University Press.

———. 1993. "Patriotism and Its Futures." *Public Culture* 5:411–29.

———. 1996. *Modernity at Large: Cultural Dimensions of Globalization*. Minneapolis: University of Minnesota Press.

———. 2000. "Grassroots Globalization and the Research Imagination." *Public Culture* 12(1): 1–19.

———. 2000. "Spectral Housing and Urban Cleansing: Notes on Millennial Mumbai." *Public Culture* 12(3): 627–51.

———. 2001. "Deep Democracy: Urban Governmentality and the Horizon of Politics." *Environment and Urbanization* 13(2): 23–43.

———. 2004. "The Capacity to Aspire: Culture and the Terms of Recognition." In *Culture and Public Action*, ed. V. Rao and M. Walton. Stanford: Stanford University Press.

Arce, A., and N. Long, eds. 1999. *Anthropology, Development, and Modernities: Exploring Discourses, Counter-tendencies, and Violence*. New York: Routledge.

Aretxaga, B. 2003. "Maddening States." *Annual Review of Anthropology* 32:393–410.

Auyero, J., and D. A. Swistun. 2009. *Flammable: Environmental Suffering in an Argentine Shantytown*. New York: Oxford University Press.

Baidya, H. R. n.d. "History of the Bagmati Civilization of the Kathmandu Valley." *The Buddha Era*, 29–32.

———. 1993. "The Glory of Tukucha, the Minor." *Kathmandu Post*. November 21.

———. 1995. "Bagmati—The Holy River." In *Images of a Century: The Changing Townscapes of the Kathmandu Valley*, by GTZ/UDLE (Deutsche Gesellschaft fur Tecchnische Zusammenarbeit and Urban Development through Local Efforts). Kathmandu: GTZ.

———. 1997. *Hail Bagmati! Guide to a Photo Exhibition*. Kathmandu: Green Energy Mission (GEM)/Nepal.

———. 1999. "What about Teku-Sankhamul Bagmati Strip?" *Kathmandu Post*, December.

———. 2001a. *He Bāgmatī, Timi āū!* (Dear Bagmati, Please Come Back!) *Kāntipur*, June 4.

———. 2001b. *Bāgmatī Sabhyatāko Khojī (Looking for the Bagmati Civilization)*. *Kāntipur* 7 Mangsir 2058.

———. 2002a. *Bāgmatī Sabhyatāko Garimā* (The Importance of the Bagmati Civilization). *Kāntipur*, December 2.

———. 2002b. *Bāgmatī Sabhyatāko Jagga Atikraman (Encroaching on Bagmati Civilization Land)*. *Kāntipur*, January 12.

———. 2002c. *Ranipokharī Bibād* (Ranipokhari Debate). *Kāntipur*, April 6.

Baidya, J. D. 2000. "Clean the Bagmati." *Kathmandu Post*, May 24.

Bajracharya, S., and L. Manandhar. 1997. *World Habitat Data Report on the Celebration of World Habitat Day*. Lalitpur: Lumati Support Group for Shelter.

Barnes, B., D. Bloor, and J. Henry. 1996. *Scientific Knowledge: A Sociological Analysis*. Chicago: University of Chicago Press.

Barnes, T., and J. Duncan, eds. 1992. *Writing Worlds: Discourse, Text, and Metaphor in the Representation of Landscape*. London: Routledge.

Baviskar, A. 1995. *In the Belly of the River: Tribal Conflicts over Development in the Narmada Valley*. New Delhi: Oxford.

———. 2003. "Between Violence and Desire: Space, Power, and Identity in the Making of Metropolitan Delhi." *International Social Science Journal* 55(175): 89–98.

———. 2005. "Toxic Citizenship: Environmental Activism, Public Interest, and

Delhi's Working Class." In *Forging Environmentalism: Justice, Livelihood, and Contested Environments*, ed. J. Bauer. New York: M. E. Sharpe.

Becker, S. 1997. "Solid Waste Management in the Kathmandu Valley." *Himalayan Research Bulletin* 17(1).

Benton, T. 1989. "Marxism and Natural Limits: An Ecological Critique and Reconstruction." *New Left Review* 178:51–86.

Bhatt, N. 2002. "King of the Jungle: An Ethnographic Study of Identity, Power, and Politics among Nepali National Park Staff." Ph.D. diss., Yale University.

Blaikie, P. 1985. *The Political Economy of Soil Erosion in Developing Countries*. New York: Longman.

Blaikie, P., and H. Brookfield. 1987. *Land Degradation and Society*. London: Methuen.

Blaikie, P., J. Cameron, and D. Seddon. 1980. *Nepal in Crisis: Growth and Stagnation at the Periphery*. Delhi: Oxford University Press.

Bledsoe, B. 2004. "Written in Stone: Inscriptions of the Kathmandu Valley's Three Kingdoms." Ph.D. diss., University of Chicago.

Bocking, S. 2004. *Nature's Experts: Science, Politics, and the Environment*. New Brunswick: Rutgers University Press.

Bourdieu, P. 1987. "Social Space and Symbolic Power." In *In Other Words: Essays toward a Reflexive Sociology*. Stanford: Stanford University Press.

———. 1990. *The Logic of Practice*. Stanford: Stanford University Press.

———. 1994. "Rethinking the State: Genesis and Structure of the Bureaucratic Field." *Sociological Theory* 12(1): 1–18.

———. 1999. "Rethinking the State: Genesis and Structure of the Bureaucratic Field." In *State/Culture: State Formation after the Cultural Turn*, ed. G. Steinmetz. Ithaca: Cornell University Press.

Boyden, S. V. 1993. "The Human Component of Ecosystems." In *Humans as Components of Ecosystems: The Ecology of Subtle Human Effects and Populated Areas*, ed. M. A. Pickett, 72–78. New York: Springer-Verlag.

Braun, B. 2005. "Environmental Issues: Writing a More Than Human Geography." *Progress in Human Geography* 29:635–50.

Brenner, N. 1997. "Global, Fragmented, Hierarchical: Henri Lefebvre's Geographies of Globalization." *Public Culture* 10(1):135–67.

———. 1998. "Global Cities, 'Glocal' States: Global City Formation and State Territorial Restructuring in Contemporary Europe." *Review of International Political Economy* 5(1): 1–37.

Brookfield, H. 1999. "A Review of Political Ecology: Issues, Epistemology, and Analytical Narratives." *Zeitschrift fur Wirtschaftsgoegraphie*, 131–47.

Brosius, J. P. 1999a. "Comments on Escobar's 'After Nature: Steps to an Antiessentialist Political Ecology.'" *Current Anthropology* 40:16–17.

———. 1999b. "Green Dots, Pink Hearts: Displacing Politics from the Malaysian Rain Forest." *American Anthropologist* 101(1): 36–57.

Brosius, J. P., and D. Russell. 2003. "Conservation from Above: An Anthropologi-

cal Perspective on Transboundary Protected Areas and Ecoregional Planning." *Journal of Sustainable Forestry* 17:39–65.
Brosius, J. P., A. Tsing, and C. Zerner. 1998. "Representing Communities: Histories and Politics of Community-Based Natural Resource Management." *Society and Natural Resources* 11:157–68.
Brundtland Commission. 1987. *Our Common Future*. New York: Oxford University Press.
Bryant, R. L. 1992. "Political Ecology: An Emerging Research Agenda in Third World Studies." *Political Geography* 11(1): 12–36.
———. 1998. "Power, Knowledge and Political Ecology in the Third World." *Progress in Physical Geography* 22:79–94.
———. 2000. "Politicized Moral Geographies: Debating Biodiversity Conservation and Ancestral Domain in the Philippines." *Political Geography* 19:673–705.
Bryant, R. L., and S. Bailey. 1997. *Third World Political Ecology*. London: Routledge.
Burghart, R. 1984. "The Formation of the Concept of Nation-State in Nepal." *Journal of Asian Studies* 44(1): 101–25.
———. 1987. "Gifts to the Gods: Power, Property, and Ceremonial in Nepal." In *Rituals of Royalty: Power and Ceremonial in Traditional Societies*, ed. David Cannadine and Simon Price, 237–70. New York: Cambridge University Press.
———. 1988. "Cultural Knowledge of Hygiene and Sanitation as a Basis for Health Development in Nepal." *Contributions to Nepalese Studies* 15(2): 185–211.
———. 1993. "His Lordship at the Cobbler's Well." In *An Anthropological Critique of Development: The Growth of Ignorance*, ed. M. Hobart, 79–99. New York: Routledge.
———. 1994. "The Political Culture of Panchayat Democracy." In *Nepal in the Nineties: Versions of the Past, Visions of the Future*, ed. Michael Hutt, 1–13. Delhi: Oxford University Press [copyright 1993].
———. 1996. *The Conditions of Listening: Essays on Religion, History, and Politics in South Asia*. New York: Oxford University Press.
Buttel, F. and P. Taylor. 1992. "How Do We Know We Have Environmental Problems?" *Geoforum* 23:405–16.
Caldeira, T. 2001. *City of Walls: Crime, Segregation, and Citizenship in São Paulo*. Berkeley: University of California Press.
Callon, M. 1995. "Four Models for the Dynamics of Science." In *Handbook of Science and Technology*, ed. S. Jasanoff, G. E. Markle, J. C. Petersen, and T. Pinch, 29–63. New York: Sage.
Campbell, B. 2003. "Resisting the Environmentalist State." In *Resistance and the State: Nepalese Experiences*, ed. D. Gellner, 83–132. Delhi: Social Science Press.
Campbell, S. 1996. "Green Cities, Growing Cities, Just Cities: Urban Planning and the Contradiction of Sustainable Development." *APA Journal* (summer): 296–312.
Campbell, T. 1992. "Socio-economic Aspects of Household Fuel Use in Pakistan." In *Natural Resources in Pakistan and Adjoining Countries: Case Studies in Applied*

Social Science, ed. M. R. Dove and C. Carpenter, 304–29. Lahore, Pakistan: Vanguard Press for the Marshall Foundation.

Caplan, L. 1975. *Administration and Politics in a Nepalese Town*. London: Oxford University Press.

Castanza, R., B. Norton, and B. Haskell. 1992. *Ecosystem Health: New Goals for Environmental Management*. Washington: Island Press.

Castells, M. 1977. *The Urban Question: A Marxist Approach*. Cambridge: MIT Press.

Cederlof, G., and K. Sivaramakrishnan, eds. 2006. *Ecological Nationalisms: Nature, Livelihoods, and Identities in South Asia*. Seattle: University of Washington Press.

Certeau, M. 2002. *The Practice of Everyday Life*. Berkeley: University of California Press.

Chakrabarty, D. 2002. "Of Garbage, Modernity and the Citizen's Gaze." In *Habitations of Modernity: Essays in the Wake of Subaltern Studies*, ed. D. Chakrabarty, 65–79. Chicago: University of Chicago Press.

Chapple, C. K. 1998. "Toward an Indigenous Indian Environmentalism." In *Purifying the Earthly Body of God: Religion and Ecology in Hindu India*, ed. L. Nelson, 13–38. Albany: State University of New York Press.

Chapple, C. K., and M. E. Tucker, eds. 2000. *Hinduism and Ecology: the Intersection of Earth, Sky, and Water*. Cambridge: Cambridge University Press.

Chatterjee, P. 1986. *Nationalist Thought and the Colonial World: A Derivative Discourse?* New Delhi: Oxford University Press.

———. 2001. "On Civil and Political Society in Post-Colonial Democracies." In *Civil Society: History and Possibilities*, ed. S. Kaviraj and S. Khilnani, 165–78. New York: Cambridge University Press.

———. 2004. *The Politics of the Governed: Reflections on Popular Politics in Most of the World*. New York: Columbia University Press.

City Diagnostic Report for City Development Strategy. 2001. Kathmandu Metropolitan City (KMC)/World Bank.

Clifford, J. 1986. "Partial Truths." In *Writing Culture: The Politics and Poetics of Ethnography*, ed. J. Clifford and G. Marcus. Berkeley: University of California Press.

———. 1988. *The Predicament of Culture: Twentieth Century Ethnography, Literature, and Art*. Cambridge: Harvard University Press.

———. 1997. "Spatial Practices: Fieldwork, Travel, and the Disciplining of Anthropology." In *Routes: Travel and Translation in the Late Twentieth Century*, 52–91. Cambridge: Harvard University Press.

Collingridge, D., and C. Reeve. 1986. *Science Speaks to Power: The Role of Experts in Policy Making*. New York: St. Martin's Press.

Collins, James P., A. Kinzig, N. Grimm, W. Fagan, D. Hope, J. Wu, and E. Borer. 2000. "A New Urban Ecology: Modeling Human Communities as Integral Parts of Ecosystems Poses Special Problems for the Development and Testing of Ecological Theory." *American Scientist* 88: 416–20.

Comaroff, J., and J. Comaroff. 2001. "Naturing the Nation: Aliens, Apocalypse and the Postcolonial State." *Journal of Southern African Studies* 27:3.

Community Led Environmental Action Network (CLEAN-Nepal). 2002. *Environmental Baseline of Bishnumati Corridor in the Kathmandu Valley*. Kathmandu: World Vision International.

Conway, D., and N. Shrestha. 1980. "Urban Growth and Urbanization in Least-Developed Countries: The Experience of Nepal, 1952-1971." *Asian Profile* 8(5).

Cooper, F., and R. Packard. 1997. Introduction. In *International Development and the Social Sciences: Essays in the History and Politics of Knowledge*, 1-41. Berkeley: University of California Press.

Corbridge, S. 2005. *Seeing the State: Governance and Governmentality in India*. Cambridge: Cambridge University Press.

Corrigan, P., and D. Sayer. 1985. *The Great Arch: English State Formation as Cultural Revolution*. Oxford: Blackwell.

Cosmos Engineering Services. 1999. *Final Report in Preparation of Conservation and Development Master Plan for Bagmati, Bishnumati-Dhobikhola Corridor*. Kathmandu: HMG Ministry of Housing and Physical Planning.

Cowen, M., and R. Shenton. 1995. "The Invention of Development." In *Power of Development*, ed. J. Crush, 27-43. London: Routledge.

Cremo, M. A., and M. Goswami. 1995. *Divine Nature: A Spiritual Perspective on the Environmental Crisis*. Los Angeles: Bhaktivedanta Press.

Creswell, T. 2005. "Moral Geographies." In *Cultural Geography: A Critical Dictionary of Key Ideas*, ed. D. Sibley, P. Jackson, D. Atkinson, and N. Washbourne. London: I. B. Tauris.

Cronon, W. 1983. *Changes in the Land: Indians, Colonists, and the Ecology of New England*. New York: Hill and Wang.

———. 1991. *Nature's Metropolis: Chicago and the Great West*. New York: W. W. Norton.

———. 1995. "The Trouble with Wilderness; or, Getting Back to the Wrong Nature." In *Uncommon Ground*, ed. W. Cronon, 69-90. New York: W. W. Norton.

Crush, J., ed. 1995. *Power of Development*. New York: Routledge.

Dahal, D. 1997. *Demolition of Squatter Houses at Kohiti and Other Environmental Problems Related to the Bishnumati Corridor: A Sociological Analysis*. Kathmandu: Asian Development Bank.

Das, P. K. 2003. "Slums: The Continuing Struggle for Housing." In *Bombay and Mumbai: The City in Transition*, ed. S. Patel and J. Masselos, 207-34. Delhi: Oxford University Press.

Davila, A. 2004. *Barrio Dreams: Puerto Ricans, Latinos, and the Neoliberal City*. Berkeley: University of California Press.

Davis, A., and J. Wagner. 2003. "Who Knows? On the Importance of Identifying 'Experts' When Researching Local Ecological Knowledge." *Human Ecology* 31(3):463-89.

Davis, D. 2000. *The Consumer Revolution in Urban China*. Berkeley: University of California Press.

Davis, M. 2004. "The Urbanization of Empire: Megacities and the Laws of Chaos." *Social Text* 81(22):4.

———. 2006. *Planet of Slums*. New York: Verso.
Dawson, A., and B. Edwards. 2004. "Introduction: Cities of the South." *Social Text* 81(22): 4.
Demeritt, D. 1994. "The Nature of Metaphors in Cultural Geography and Environmental History." *Progress in Human Geography* 18:163.
Dimitrov, R. S. 2003. "Knowledge, Power, and Interests in Environmental Regime Formation." *International Studies Quarterly* 47(1): 123–50.
Dirks, N. 1990. "The Original Caste." In *India through Hindu Categories*, ed. M. Marriott, 59–77. New Delhi: Sage.
Dixit, K. M. 1995. "A Valley Changing." In *Images of a Century: The Changing Townscapes of the Kathmandu Valley*. Kathmandu: Deutsche Gesellschaft fur Technische Zusammenarbeit and Urban Development through Local Efforts.
———. 1997. "Foreign Aid in Nepal: No Bang for the Buck." *Studies in Nepali History and Society* 2(1): 173–86.
Dixit, K. M., and S. Ramachandran, eds. 2002. *State of Nepal*. Lalitpur: Himal Books.
Douglas, M. 1966. *Purity and Danger: An Analysis of the Concepts of Pollution and Taboo*. New York: Routledge & Kegan Paul PLC.
———. 1986. *How Institutions Think*. New York: Syracuse University Press.
Dove, M. R. 1993. "Uncertainty, Humility, and Adaptation to the Tropical Forest: The Agricultural Augury of the Kantu." *Ethnology* 32(2): 145–67.
———. 1998a. "Local Dimensions of 'Global' Environmental Debates." In *Environmental Movements in Asia*, ed. A. Kalland and G. Persoon. Man and Nature Series. Surrey: Curzon.
———. 1998b. "Representations of the 'Other' by Others: The Ethnographic Challenge Posed by Planters' View of Peasants in Indonesia." In *Agrarian Transformation in the Indonesian Uplands*, ed. T. Li. Amsterdam: Harwood Academic.
———. 1999. "Writing for, Versus about, the Ethnographic Other: Issues of Engagement and Reflexivity in Working with a Tribal NGO in Indonesia." *Identities* 6(2–3): 225–53.
———. 2000. "Inter-Disciplinary Borrowing in Ecological Anthropology, and the Critique of Modern Science." In *New Directions in Anthropology and Environment*, ed. C. L. Crumley. Walnut Creek, Calif.: AltaMira.
———. 2003. "Forest Discourses in South and Southeast Asia: A Comparison with Global Discourses." In *Nature in the Global South: Environmental Projects in South and Southeast Asia*, ed. P. Greenough and A. Tsing. Durham: Duke University Press.
Dove, M. R., and M. H. Kahn. 1995. "Competing Constructions of Calamity: The Case of the May 1991 Bangladesh Cyclone." *Population and Environment* 16(5): 445–71.
———. 2000. "Resources In, Information Out: Asymmetries in Government and Their Implications for Sustainable Development." *World Development* 28.
Downey, G. L., and J. Dumit. 1997. *Cyborgs and Citadels: Anthropological Interven-*

tions in Emerging Sciences and Technologies. Santa Fe, N.M.: School of American Research Press.

Driver, F. 1988. "Moral Geographies: Social Science and the Urban Environment in Mid-Nineteenth-Century England." *Transactions of the Institute of British Geographers* 13:275–87.

Drury, W. H., and I. Nisbet. 1973. "Succession." *Journal of the Arnold Arboretum* 4:221–368.

Dumit, J. 2004. *Picturing Personhood: Brain Scans and Biomedical Identity.* Princeton: Princeton University Press.

Dumont, L. 1970. *Homo Hierarchicus: An Essay on the Caste System*, trans. Mark Sainsbury. Chicago: University of Chicago Press.

Eckholm, E. 1976. *Losing Ground: Environmental Stress and World Food Prospects.* New York: W. W. Norton.

Ehrlich, P. 1968. *Population Bomb.* New York: Ballantine.

Eisenstadt, S. N. 2000. "Multiple Modernities." *Daedalus* 129(1): 1–29.

Emmel, N. D., and J. G. Soussan. 2001. "Interpreting Environmental Degradation and Development in the Slums of Mumbai, India." *Land Degradation and Development* 12:277–83.

English, R. 1985. "Himalayan State Formation and the Impact of British Rule in the Nineteenth Century." *Mountain Research and Development* 5(1): 61–78.

Escobar, Arturo. 1992. "Imagining a Post-Development Era? Critical Thought, Development, and Social Movements." *Social Text* 31/32:20–56.

———. 1994. *Encountering Development.* Princeton: Princeton University Press.

Evans, P. 1979. *Dependent Development: The Alliance of Multinational, State, and Local Capital in Brazil.* Princeton: Princeton University Press.

Evans, P., ed. 2002. *Livable Cities? Urban Struggles for Livelihood and Sustainability.* Berkeley: University of California Press.

Evans, P., D. Rueschemeyer, and T. Skocpol, eds. 1985. *Bringing the State Back In.* London: Cambridge University Press.

Fairhead, J., and M. Leach. 1996. *Misreading the African Landscape: Society and Ecology in Forest-Savanna Mosaic.* Cambridge: Cambridge University Press.

Faure, D., and H. F. Siu. 1995. *Down to Earth: The Territorial Bond in South China.* Stanford: Stanford University Press.

Federation of Canadian Municipalities, AMC Earth and Environmental Limited, and Nepalconsult. 2001. "Achieving Environmental and Community Sustainability through Integrated Water Resource Management and Strengthening of Municipal Government in Nepal: Water Optimization Pilot Project, Kathmandu Valley, Nepal." Position paper submitted to Ministry of Physical Planning and Works and Municipal Association of Nepal, October.

Feldhaus, A. 1995. *Water and Womanhood: Religious Meanings of Rivers in Maharashtra.* New York: Oxford University Press.

Ferguson, J. 1994. *Anti-Politics Machine: "Development," Depoliticization, and Bureaucratic Power in Lesotho.* Minneapolis: University of Minnesota Press.

———. 1999. *Expectations of Modernity: Myths and Meanings of Urban Life on the Zambian Copperbelt*. Berkeley: University of California Press.

———. 2004. "Power Topographies." In *A Companion to the Anthropology of Politics*, ed. D. Nugent and J. Vincent. Malden, Mass.: Blackwell.

Ferguson, J., and A. Gupta. 2002. "Spatializing States: Toward an Ethnography of Neoliberal Governmentality." *American Ethnologist* 29:981–1002.

Fernandez, K., ed. 1997. "Evictions in Nepal 1996." In *Eviction Watch Asia*, 47–50. Singapore: Asian Coalition for Housing Rights.

Fisher, D. R., and W. R. Freudenburg. 2001. "Ecological Modernization and Its Critics: Assessing the Past and Looking toward the Future." *Society and Natural Resources* 14:701–9.

Fisher, J. 2000. *Living Martyrs: Individuals and Revolution in Nepal*. Delhi: Oxford University Press.

Forbes, A. 1999. "The Importance of Being Local: Villagers, NGOs, and the World Bank in the Arun Valley, Nepal." *Identities* 6(2–3): 319–44.

Franklin, S. 1997. "Autonomous Agents as Embodied AI." *Cybernetics and Systems* 28:499–520.

Fujikura, T. 1996. "Technologies of Improvement, Locations of Culture: American Discourses of Democracy and 'Community Development' in Nepal." *Studies in Nepali History and Society* 1(2): 271–311.

———. 2001. "Discourses of Awareness: Notes for a Criticism of Development in Nepal." *Studies in Nepali History and Society* 6(2): 271–313.

Gallagher, K. 1991. "An Exploration into the Causes of Squatting in the Kathmandu Valley." Ph.D. diss., Tribhuvan University, Kathmandu.

Gandy, M. 2002. *Concrete and Clay: Reworking Nature in New York City. Urban and Industrial Environments*. Cambridge: MIT Press.

———. 2006. "Zones of Indistinction: Bio-Political Contestations in the Urban Arena." *Cultural Geographies* 13:497–516.

———. 2009. "The Uses of Max Weber: Legitimation and Amnesia in Buddhology, South Asian History, and Anthropological Practice Theory." In *The Oxford Handbook of the Sociology of Religion*, ed. P. Clarke, 48–62. Oxford: Oxford University Press.

Gellner, D. 1997. *Caste Communalism and Communism: Newars and the Nepalese State, in Nationalism and Ethnicity in a Hindu Kingdom*, ed. D. Gellner, J. Pfaff-Czarnecka, and J. Whelpton, 151–84. Amsterdam: Harwood Academic Publishers.

———. 2001. "Does Symbolism 'Construct an Urban Mesocosm'? Robert Levy's Mesocosm and the Question of Value Consensus in Bhaktapur." In *The Anthropology of Buddhism and Hinduism: Weberian Themes*, ed. D. Gellner, 293–315. New Delhi: Oxford University Press.

———. 2002. Introduction. *Transformations of the Nepalese State, in Resistance and the State: Nepalese Experiences*, ed. D. Gellner, 1–30. New Delhi: Social Science Press.

———. 2007a. "Caste, Ethnicity, and Inequality in Nepal." *EPW* 42(20):1823–28.
———. 2007b. "Democracy in Nepal: Four Models." *Seminar* 576:50–56.
Gellner, D., ed. 2003. *Resistance and the State: Nepalese Experiences*. Delhi: Social Science Press.
Gellner, D., and K. Hachhethu, eds. 2008. *Local Democracy in South Asia: Microprocesses of Democratization in Nepal and Its Neighbors*. New Delhi: Sage.
Gellner, D., J. Pfaff-Czarnecka, and J. Whelpton, eds. 1997. *Nationalism and Ethnicity in a Hindu Kingdom: The Politics of Culture in Contemporary Nepal*. Amsterdam: Harwood Academic.
Gellner, D., and M. B. Karki. 2008. "Democracy and Ethnic Organizations in Nepal." In *Local Democracy in South Asia: Microprocesses of Democratization in Nepal and Its Neighbors*, ed. D. Gellner and K. Hachhethu, 105–27. New Delhi: Sage.
Ghannam, F. 2002. *Remaking the Modern: Space, Relocation, and the Politics of Identity in a Global Cairo*. Berkeley: University of California Press.
Gold, A., and B. R. Gujar. 2002. *In the Time of Trees and Sorrows: Nature, Power, and Memory in Rajasthan*. Durham: Duke University Press.
Gooding, D. 1992. "Putting Agency Back into Experiment." In *Science as Practice and Culture*, ed. A. Pickering, 65–112. Chicago: University of Chicago Press.
Gray, J. 2006. *Domestic Mandala: Architecture of Lifeworlds in Nepal*. Burlington, Vt.: Ashgate.
Greenough, P. 2003. "Pathogens, Pugmarks and Political Emergency: The 1970s South Asian Debate on Nature." In *Nature in the Global South: Environmental Projects in South and Southeast Asia*, ed. P. Greenough and A. L. Tsing, 201–30. Durham: Duke University Press.
Greenough, P., and A. Tsing, eds. 2003. *Nature in the Global South: Environmental Projects in South and Southeast Asia*. Durham: Duke University Press.
Gregory, S. 1998. *Black Corona: Race and the Politics of Place in an Urban Community*. Princeton: Princeton University Press.
Grimm, N. B., L. J. Baker, and D. Hope. 2003. "An Ecosystem Approach to Understanding Cities: Familiar Foundations and Uncharted Frontiers." In *Understanding Urban Ecosystems: A New Frontier for Science and Education*, ed. A. R. Berkowitz, C. H. Nilon, and K. S. Hollweg, 95–114. New York: Springer.
Grove, R. H. 1989. *Green Imperialism: Colonial Expansion, Tropical Island Edens and the Origins of Environmentalism 1600–1860*. Cambridge: Cambridge University Press.
Grove-White, R. 1993. "Environmentalism: A New Moral Discourse for Technological Society?" In *Environmentalism: The View from Anthropology*, ed. K. Milton. New York: Routledge.
GTZ/UDLE (Deutsche Gesellschaft fur Tecchnische Zusammenarbeit and Urban Development through Local Efforts). 1995. *Images of a Century: The Changing Townscapes of the Kathmandu Valley*. Kathmandu: GTZ.
Guneratne, A. 2002. *Many Tongues, One People: The Making of Tharu Identity in Nepal*. Ithaca: Cornell University Press.

Gupta, A. 1995. "Blurred Boundaries: The Discourse of Corruption, the Culture of Politics, and the Imagined State." *American Ethnologist* 22(2): 375–402.

Gupta, A., and J. Ferguson. 1992. "Beyond 'Culture': Space, Identity, and the Politics of Difference." *Cultural Anthropology* 7:6–23.

———. 1997a. "Discipline and Practice: 'The Field' as Site, Method and Location in Anthropology." In *Anthropological Locations: Boundaries and Grounds of a Field Science*, ed. A. Gupta and J. Ferguson, 1–46. Berkeley: University of California Press.

———. 1997b. "Culture, Power, Place. Ethnography at the End of an Era." In *Culture, Power, Place: Explorations in Critical Anthropology*, ed. A.Gupta and J. Ferguson. Durham: Duke University Press.

Guthman, J. 1997. "Representing Crisis: The Theory of Himalayan Environmental Degradation and the Project of Development in Post-Rana Nepal." *Development and Change* 28:45–69.

Gutschow, N. 1985. *Heritage of the Kathmandu Valley: Proceedings of an International Conference*. St. Augustine, Fla.: VGH Wissenschaftsberlag.

Gutschow, N., A. Michaels, C. Ramble, and E. Steinkellner. 2003. *Sacred Landscape of the Himalaya: Proceedings of an International Conference at Heidelberg*. Vienna: Austrian Academy of Sciences Press.

Gyawali, D. 1992. "Stress, Strain, and Insults." *Himal* 5:9–13.

———. 2001. *Water in Nepal*. Kathmandu: Himal.

Hachhethu, K. 2002. *Party Building in Nepal: Organization, Leadership and People*. Kathmandu: Mandala Book Point.

Hada, J. 2001. "Housing and Squatter Settlements." In *City Diagnostic Report for City Development Strategy*. Kathmandu Metropolitan City (KMC)/World Bank.

Hall, P. 1988. *Cities of Tomorrow: An Intellectual History of Urban Planning and Design in the Twentieth Century*. Cambridge: Basil Blackwell.

Hall, S. 1996. In *Critical Dialogues in Cultural Studies*, ed. David Morley, David, Stuart Hall, and Kuan-Hsing Chen. London: Routledge.

Hamilton, F. 1819. *An Account of the Kingdom of Nepal, and of Territories and of the Territories Annexed to This Dominion by the House of Gorkha*. Edinburgh: A Constable Press.

Hangen, S. 2000. "Making Mongols: Ethnic Politics and Emerging Identities in Nepal." Ph.D. diss., University of Wisconsin, Madison.

———. 2007. "Creating a 'New Nepal': The Ethnic Dimension." *Policy Studies* 34. Washington: The East-West Center.

———. 2010. *The Rise of Ethnic Politics in Nepal: Democracy in the Margins*. New York: Routledge.

Hannerz, U. 1987. "The World in Creolization." *Africa* 57:546–59.

———. 1996. *Transnational Connections*. New York: Routledge.

Hansen, T. B. 2001. *Wages of Violence: Naming and Identity in Postcolonial Bombay*. Princeton: Princeton University Press.

Hansen, T. B., and F. Stepputat. 2001. *States of Imagination: Ethnographic Explorations of the Postcolonial State*. Durham: Duke University Press.

Haraway, D. J. 1989. *Primate Visions: Gender, Race, and Nature in the World of Modern Science*. New York: Routledge.
———. 1991. *Simians, Cyborgs, and Women: The Reinvention of Nature*. New York: Routledge.
———. 1997. *Modest Witness@Second Millennium. FemaleMan Meets OncoMouse: Feminism and Technoscience*. New York: Routledge.
Hardoy, J. E., and D. Satterthwaite. 1989. *Squatter Citizen: Life in the Urban Third World*. London: Earthscan.
Hart, G. P. 2002. *Disabling Globalization: Places of Power in Post-Apartheid South Africa*. Berkeley: University of California Press.
Harvey, D. 1990. *The Condition of Postmodernity: An Enquiry into the Origins of Cultural Change*. Cambridge: Blackwell.
———. 2008. "The Right to the City." *New Left Review* 53.
Hayden, D. 1995. *The Power of Place: Urban Landscapes and Public History*. Cambridge: MIT Press.
Heesterman, J. C. 1985. *The Inner Conflicts of Tradition*. Chicago: University of Chicago Press.
Herzfeld, M. 2006. "Spatial Cleansing: Monumental Vacuity and the Idea of the West." *Journal of Material Culture* 11(1–2): 127–49.
His Majesty's Government of Nepal (HMG). 1996. *National Shelter Policy*. Kathmandu: HMG.
His Majesty's Government of Nepal (HMG) and Asian Development Bank (ADB). 1991. *Environmental Policy Assessment: Kathmandu Urban Development Plans and Programs: Concept Plan for the Bishnumati Corridor*. Kathmandu: Halcrow Fox and Associates.
———. 1992. *Project Kathmandu Urban Development Preparation Report*. Kathmandu: HMG.
His Majesty's Government of Nepal (HMG), Ministry of Housing and Physical Planning and Kathmandu Urban Development Project. 1996. *Community Participation in Development Programs: Working Paper*. Kathmandu: COWI Consult/Padco/Nepal Consult.
His Majesty's Government of Nepal (HMG), National Habitat Committee. 2001. *Nepal National Report: United Nations General Assembly on Human Settlements Istanbul +5 Meeting*. Kathmandu: HMG.
Hjortshoj, K. 1979. *Urban Structures and Transformations in Lucknow, India*. Ithaca: Cornell University Program in International Studies in Planning.
Hobsbawm, E., and T. Ranger, 1983. *The Invention of Tradition*. Cambridge: Cambridge University Press.
Hoeck, B. 1990. "Does Divinity Protect the King? Ritual and Politics in Nepal." *Contributions to Nepalese Studies* 17(2): 147–55.
Hofer, A. 1979. *The Caste Hierarchy and the State of Nepal: A Study of the Muluki Ain of 1854*. Innsbruck: Universtatsverlag Wagner.
Hoftun, M. 1993. "The Dynamics and Chronology of the 1990 Revolution." In *Nepal in the Nineties*, ed. Michael Hutt. Delhi: Oxford University Press.

Hoftun, M., and W. Raeper. 1992. *Spring Awakening: An Account of the 1990 Revolution in Nepal*. Delhi: Viking.

Hoftun, M., W. Raeper, and J. Whelpton. 1999. *People, Politics and Ideology: Democracy and Social Change in Nepal*. Kathmandu: Mandala Book Point.

Hogle, L. 1995. "Standardization across Non-Standard Domains: The Case of Organ Procurement." *Science, Technology, and Human Values* 20(4): 482–500.

Holling, C. S., P. Taylor, and M. Thompson. 1991. "From Newton's Sleep to Blake's Fourfold Vision: Why the Climax Community and the Rational Bureaucracy Are Not the Ends of the Ecological and Socio-Cultural Roads." *Annals of Earth* 9(3): 19–21.

Holston, J. 1989. *The Modernist City: An Anthropological Critique of Brasilia*. Chicago: University of Chicago Press.

Holston, J., and A. Appadurai. 1999. *Cities and Citizenship*. Durham: Duke University Press.

Howard, E. 1965 [1898]. *Garden Cities of To-Morrow*. Cambridge: MIT Press.

Hunt, J. 2001. *Situation Analysis Report: A Study of Poor Urban Communities in the Kathmandu Valley*. Kathmandu: Action Aid/Lumanti.

Hutt, M. 1995. *Nepal: A Guide to the Art and Architecture of the Kathmandu Valley*. Gartmore, U.K.: Kiscadale.

———. 2004. *Himalayan People's War: Nepal's Maoist Rebellion*. Bloomington: Indiana University Press.

Ifeka, C. 1987. "Domestic Space as Ideology in Goa." *Contributions to Indian Sociology* 21:307–29.

Inden, R. 1978. *Ritual, Authority and Cyclic Time in the Hindu Kingship*. In *Kingship and Authority in South Asia*, ed. J. F. Richards, 28–73. Madison: University of Wisconsin Press.

INSEC. 1992. *Bonded Labor in Nepal: Under Kamaiya System*. Kathmandu: INSEC.

Ishii, H., D. Gellner, and K. Nawa. 2007. *Political and Social Transformations in North India and Nepal*. Delhi: Manohar.

Ives, J., and B. Messerli. 1989. *The Himalayan Dilemma: Reconciling Development and Conservation*. London: Routledge.

Jasanoff, S. 2004. *States of Knowledge: The Coproduction of Science and the Social Order*. London: Routledge.

Jasanoff, S., and M. L. Martello, eds. 2004. *Earthly Politics: Local and Global in Environmental Governance*. Cambridge: MIT Press.

Jessop, B. 1990. *State Theory: Putting the Capitalist State in its Place*. Cambridge: Polity Press.

Joseph, G., and D. Nugent. 1994. *Everyday Forms of State Formation: Revolution and the Negotiation of Rule in Modern Mexico*. Durham: Duke University Press.

Joyce, P. 2003. *The Rule of Freedom: Liberalism and the Modern City*. New York: Verso.

Kaika, M. 2005. *City of Flows: Modernity, Nature, and the City*. London: Routledge.

Kapferer, B. 1988. *Legends of People, Myths of State: Violence, Intolerance, and Political Culture in Sri Lanka and Australia. Washington: Smithsonian Institution Press.

Kathmandu Metropolitan City. 2001. *Diagnostic Report for City Development Strategy*. Kathmandu: KMC/World Bank.

KC, B. K. 1998. "Trends, Patterns, and Implications of Rural-to-Urban Migration in Nepal. Kathmandu." Central Department of Population Studies, Tribhuvan University.

Kendall, L. 1998. "Who Speaks for Korean Shamans when Shamans Speak of the Nation?" In *Making Majorities: Constituting the Nation in Japan, Korea, China, Malaysia, Fiji, Turkey, and the United States*, ed. D. Gladney, 55–72. Stanford: Stanford University Press.

Knorr-Cetina, K. D. 1981. *The Manufacture of Knowledge: An Essay on the Constructivist and Contextual Nature of Science*. Oxford: Pergamon.

Kramrisch, S. 1976. *The Hindu Temple*. 2 Volumes. Delhi: Motilal Banarsidas.

Lakier, G. 2005. "The Myth of the State Is Real: Notes on the Study of the State in Nepal." *Studies in Nepali History and Society* 10(1): 135–70.

Latour, B. 1993. *We Have Never Been Modern*, trans. Catherine Porter. Cambridge: Harvard University Press.

———. 1999. *Pandora's Hope: Essays on the Reality of Science Studies*. Cambridge: Harvard University Press.

———. 2004. *Politics of Nature: How to Bring the Sciences into Democracy*. Cambridge: Harvard University Press.

———. 2005. "From Realpolitik to Dingpolitik or How to Make Things Public." In *Making Things Public: Atmospheres of Democracy*, ed. B. Latour and P. Weibel, 14–41. Cambridge: MIT Press.

Latour, B., and S. Woolgar. 1986. *Laboratory Life: The Social Construction of Scientific Facts*. Princeton: Princeton University Press.

Lawoti, Mahendra, ed. 2007. *Contentious Politics and Democratization in Nepal: The Maoist Insurgency, Identity Politics, and Social Mobilization since 1990*. Delhi: Sage.

Lefebvre, H. 1991. *The Production of Space*. Cambridge: Blackwell.

———. 2003 [1970]. *The Urban Revolution*. Minneapolis: University of Minnesota Press.

Levine, N. 1987. "Caste, State, and Ethnic Boundaries in Nepal." *Journal of Asian Studies* 46(1): 71–88.

Levy, R. 1990. *Mesocosm: Hinduism and the Organization of a Traditional Newar City in Nepal*. Berkeley: University of California Press.

Li, T., 1999. "Compromising Power: Development, Culture, and Rule in Indonesia." *Cultural Anthropology* 14(3): 295–322.

Liechty, M. 1996. "Kathmandu as Translocality: Multiple Places in a Nepali Space." In *The Geography of Identity*, ed. P. Yaeger. Ann Arbor: University of Michigan Press.

———. 1997. "Selective Exclusion: Foreigners, Foreign Goods, and Foreignness in Modern Nepali History." *Studies in Nepali History and Society* 2(1): 5–68.

———. 2003. *Suitably Modern: Making Middle Class Culture in a New Consumer Society.* Princeton: Princeton University Press.
Lorenz, E. 1996 [1961]. *The Essence of Chaos.* The Jessie and John Danz Lecture Series. Seattle: University of Washington Press.
Low, S., ed. 1999. *Theorizing the City: The New Urban Anthropology Reader.* New Brunswick: Rutgers University Press.
———. 2003. *Behind the Gates: Life, Security, and the Pursuit of Happiness in Fortress America.* New York: Routledge.
Lowe, P. 2001. *Kamaiya: Slavery and Freedom in Nepal.* Kathmandu: Mandala Book Point.
Ludden, D. 1992. "India's Development Regime." In *Colonialism and Culture*, ed. N. Dirks, 247–88. Ann Arbor: University of Michigan Press.
Lumanti Support Group for Shelter. 1998. *CityCare Annual Newsletter of Lumanti Support Group for Shelter.* No. 1. Kathmandu: Lumanti.
———. 2000. *Solid Waste in Kathmandu: Problems and Management.* Report.
———. 2002a. *Report on Evictions at Thapatali.* Kathmandu: Lumanti.
———. 2002b. *CityCare Annual Newsletter of Lumanti Support Group for Shelter.* Kathmandu: Lumanti.
———. 2006. *New Beginnings: Housing the Urban Poor: A Case Study of Kirtipur Housing Project.* Kathmandu: Lumanti.
Lynch, M. 1985. *Art and Artifact in Laboratory Science: A Study of Shop Work and Shop Talk in the Laboratory.* London: Routledge and Kegan Paul.
MacFayden, J. T., and J. W. Vogt. 1977. "The City as Mandala: Bhaktapur." *Ekistics* 44 (265): 307–12.
Machlis, G. E., J. E. Force, and W. R. Burch, Jr. 1997. "The Human Ecosystem Part I: The Human Ecosystem as an Organizing Concept in Ecosystem Management." *Society and Natural Resources* 10:347–67.
Main, H., and S. W. Williams, eds. 1994. *Environment and Housing in Third World Cities.* Chichester: John Wiley.
Malkki, L. 1992. "National Geographic: The Rooting of Peoples and the Territorialization of National Identity among Scholars and Refugees." *Cultural Anthropology* 7(1): 24–44.
Manandhar, R. 2001. "Maitighar Corner to Get Exotic Look with Mandala, Stupa, and Water Spouts." *Kathmandu Post*, December 27.
Mankekar, P. 1999. *Screening Culture, Viewing Politics: An Ethnography of Television, Womanhood, and Nation in Postcolonial India.* Durham: Duke University Press.
———. 2004. "Dangerous Desires: Television and Erotics in Late Twentieth-Century India." *Journal of Asian Studies* 63:403–31.
Marcus, G. 1995. "Ethnography in/of the World System: The Emergence of Multi-Sited Ethnography." *Annual Review of Anthropology* 24:95–117.
———. 1998. *Ethnography through Thick and Thin.* Princeton: Princeton University Press.
Marcus, G., and M. Fisher. 1986. *Anthropology as Cultural Critique: An Experimental Moment in the Human Sciences.* Chicago: University of Chicago Press.

Marcuse, P., and R. van Kempen. 2002. *Of States and Cities: The Partitioning of Urban Space*. New York: Oxford University Press.

Marzluff, J. 2008. *Urban Ecology: An International Perspective on the Interaction between Humans and Nature*. New York: Springer.

Mawdsley, E. 2005. "The Abuse of Religion and Ecology: The Vishva Hindu Parishad and Tehri Dam." *Worldviews: Environment, Culture, Religion* 8(2):1–24.

———. 2006. "Hindu Nationalism, Postcolonialism, and Environmental Discourses in India." *Geoforum* 37(3): 380–90.

May, R. 1974. "Biological Populations with Nonoverlapping Generations: Stable Points, Stable Cycles, and Chaos." *Science* 186 (November 15): 645–47.

Mayer, M. 2000. "Urban Social Movements in an Era of Globalization." In *Urban Movements in a Globalizing World*, ed. P. Hamel, H. Lustiger-Thaler, and M. Mayer, 141–57. London: Routledge.

Mazzarella, W. 2003. *Shoveling Smoke: Advertising and Globalization in Contemporary India*. Durham: Duke University Press.

Mbembe, A. 2001. *On the Postcolony*. Berkeley: University of California Press.

McDonnell, M., and S. Pickett. 1990. "Ecosystem Structure and Function along Urban-Rural Gradients: An Unexploited Opportunity for Ecology." *Ecology* 71:1232–37.

McKinney, M. L. 2002. "Urbanization, Biodiversity, and Conservation." *BioScience* 52:883–90.

Mehta, S. 2004. *Maximum City: Bombay Lost and Found*. New York: Alfred A. Knopf.

Metcalf and Eddy in association with CEMAT Consultants. 2000. *Urban Water Supply Reforms in the Kathmandu Valley: Wastewater Management Plan Assessment*. Kathmandu: CEMAT Consultants.

Metz, J. 2010. "Downward Spiral? Interrogating Narratives of Environmental Change in the Himalaya." In *Symbolic Ecologies: Nature and Society in the Himalaya*, ed. Arjun Guneratne. New York: Routledge.

Michael, B. 1999. "Statemaking and Space on the Margins of Empire: Rethinking the Anglo-Gorkha War of 1814–1816." *Studies in Nepali History and Society* 14(2): 247–94.

Migdal, J., A. Kohl, and V. Shue. 1994. *State Power and Social Forces: Domination and Transformation in the Third World*. Cambridge: Cambridge University Press.

Mitchell, T. 1990. "Everyday Metaphors of Power." *Theory and Society* 19:545–77.

———. 1991a. "America's Egypt: Discourse in the Development Industry." *Middle East Report* 169 (March–April): 18–34, 36.

———. 1991b. "The Limits of the State: Beyond Statist Approaches and their Critics." *American Political Science Review* 85(1): 77–96.

———. 1999. "Rethinking the State: Genesis and Structure of the Bureaucratic Field." In *State/Culture: State Formation after the Cultural Turn*, ed. G. Steinmetz. Ithaca: Cornell University Press.

———. 2002. *Rule of Experts: Egypt, Techno-Politics, Modernity*. Berkeley: University of California Press.

Mitchell, T., ed. 2000. *Questions of Modernity*. Minneapolis: University of Minnesota Press.

Moffat, T., and E. Finnis. 2005. "Considering Social and Material Resources: The Political Ecology of a Peri-Urban Squatter Community in Nepal." *Habitat International* 29(3): 453–68.

Moore, D. 1994. "Contesting Terrain in Zimbabwe's Eastern Highlands: Political Ecology, Ethnography, and Peasant Resource Struggles." *Economic Geography* 70:380–401.

———. 1998. "Clear Waters and Muddied Histories: Environmental History and the Politics of Community in Zimbabwe's Eastern Highlands." *Journal of Southern African Studies* 24(2): 377–403.

Moore, D., J. Kosek, and A. Pandian, eds. 2003. *Race, Nature, and the Politics of Difference*. Durham: Duke University Press.

Moore, M. 1990. "The Kerala House as a Hindu Cosmos." In *India through Hindu Categories*, ed. M. Marriott, 169–202. New Delhi: Sage.

Mosse, D. 1997. "The Symbolic Making of a Common Property Resource: History, Ecology, and Locality in a Tank-Irrigated Landscape in South India." *Development and Change* 28:467–504.

———. 2005. *Cultivating Development: An Ethnography of Aid Policy and Practice*. Ann Arbor: Pluto Press.

Mosse, D., and M. Sivan. 2003. *The Rule of Water: Statecraft, Ecology, and Collective Action in South India*. New Delhi: Oxford University Press.

Munch, C., and C. Flyen. 1990. *Historic City Core of Kathmandu: Changes and Upgrading of Public Areas*. Trondheim: Norwegian Institute of Technology Division of Town and Regional Planning.

Muni, S. 2002. "SAARC at Crossroads." *Himal South Asian* 15(1).

Nandy, A. 1998. "Introduction: Indian Popular Cinema as the Slum's Eye View of Politics." In *The Secret Politics of Our Desires: Innocence, Culpability, and Indian Popular Cinema*, ed. A. Nandy. Delhi: Oxford University Press.

Narayanan, V. 2001. "Water, Wood, and Wisdom: Ecological Perspectives from the Hindu Traditions." *Daedalus* 130(4): 179–206.

National Habitat Committee Nepal. 2001. *Nepal National Report to Istanbul +5*. Kathmandu: MHGN.

Nelson, L. E., ed. 1998. *Purifying the Earthly Body of God: Religion and Ecology in Hindu India*. L. Nelson, ed. Albany: State University of New York Press.

Nepal Environmental and Scientific Services. 1997. *Final Report: Environmental Study of the Bagmati Watershed and Mitigation of River Pollution*. Kathmandu: HMG Ministry of Water Resources.

Neuwirth, R. 2006. *Shadow Cities: A Billion Squatters, a New Urban World*. New York: Routledge.

Norval, A. 2001. "Reconstructing National Identity and Renegotiating Memory:

The Work of the TRC." In *States of Imagination: Ethnographic Explorations of the Postcolonial State*, ed. T. B. Hansen and F. Stepputat. Durham: Duke University Press.

Nugent, D. 1994. "Building the State, Making the Nation: The Bases and Limits of State Centralization in 'Modern' Peru." *American Anthropologist* 96(2): 333–69.

———. 2004. "Governing States." In *A Companion to the Anthropology of Politics*, ed. D. Nugent and J. Vincent. Malden, Mass: Blackwell.

Nuijten, M. 1998. "In the Name of the Land: Organization, Transnationalism, and the Culture of the State in a Mexican Ejido." Ph.D. diss., University of Wageningen, Holland.

O'Connor, J. 1989. "Capitalism, Nature, Socialism: A Theoretical Introduction." *Capitalism, Nature, Socialism* 1(1): 11–38.

O'Dougherty, M. 2002. *Consumption Intensified: The Politics of Middle Class Daily Life in Brazil*. Durham: Duke University Press.

Odum, E. 1953. *Fundamentals of Ecology*. Philadelphia: University of Pennsylvania Press.

Ogura, K. 2008. "Maoist People's Governments, 2001–2005: The Power in Wartime." In *Local Democracy in South Asia: Microprocesses of Democratization in Nepal and Its Neighbors*, ed. D. Gellner and K. Hachhethu. New Delhi: Sage.

Ong, A. 1999. *Flexible Citizenship: The Cultural Logics of Transnationality*. Durham: Duke University Press.

Onta, P. 1994. "Rich Possibilities: Notes on Social History in Nepal," *Contributions to Nepalese Studies* 21(1): 1–43.

———. 1996. "Creating a Brave New Nation in British India: The Rhetoric of Jati Improvement, Rediscovery of Bhanubhakta and the Writing of Bir History." *Studies in Nepali History and Society* 1(1): 37–76.

———. 1997. "Activities in a 'Fossil State': Balkrishna Sama and the Improvisation of Nepali Identity." *Studies in Nepali History and Society* 2(1): 69–102.

———. 2002. "Critiquing the Media Boom." In *State of Nepal*, ed. K. M. Dixit and S. Ramachandaran, 253–69. Lalitpur: Himal Books.

Parlange, M. 1998. "The City as Ecosystem." *Bioscience* 48:581–82.

Peet, R., and M. Watts, eds. 1996. *Liberation Ecologies: Environment, Development, Social Movements*. New York: Routledge.

Pelling, M. 1999. "The Political Ecology of Flood Hazard in Urban Guyana." *Geoforum* 30(3): 249–61.

Pemberton, J. 1994. *On the Subject of Java*. Ithaca: Cornell University Press.

Peters, P. 1996. "Introduction: Who's Local Here? The Politics of Participation in Development." *Cultural Survival Quarterly* 20(3).

Pfaff-Czarnecka, J. 1997. "Vestiges and Visions: Cultural Change in the Process of Nation-Building in Nepal." In *Nationalism and Ethnicity in a Hindu Kingdom: The Politics of Culture in Contemporary Nepal*, ed. D. Gellner, Pfaff-Czarnecka, and J. Whelpton, 419–70. Amsterdam: Harwood Academic.

Pickett, S. T. A. 1997. "A Conceptual Framework for the Study of Human Ecosystems in Urban Areas." *Urban Ecosystems* 1:185–92.

Pickett, S. T. A., and M. L. Cadenasso. 2002. "The Ecosystem as a Multidimensional Concept: Meaning, Model, and Metaphor." *Ecosystems* 5:1–10.

Pickett, S. T. A., M. L. Cadenasso, J. M. Grove, C. H. Nilon, R. V. Pouyat, W. C. Zipperer, and R. Costanza. 2001. "Urban Ecological Systems: Linking Terrestrial Ecological, Physical, and Socioeconomic Components of Metropolitan Areas." *Annual Review of Ecology and Systematics* 32:127–57.

Pigg, S. 1992. "Inventing Social Categories through Place: Social Representations and Development in Nepal." *Comparative Studies in Society and History* 34(3): 492–511.

———. 1996. "The Credible and the Credulous: The Question of 'Villager's Beliefs' in Nepal." *Cultural Anthropology* 11(2): 160–202.

Pradhan, B. 1998. "Water Quality Assessment of the Bagmati River and Its Tributaries." Ph.D. diss., University of Agricultural Sciences (BOKU), Vienna.

Pradhan, R. 2002. "Ethnicity, Caste, and Pluralistic Society." In *State of Nepal*, ed. K. M. Dixit and S. Ramachandaran, 1–21. Kathmandu: Himal Books.

Proksch, A. 1995. Foreword. In *Images of a Century: The Changing Townscapes of the Kathmandu Valley*. GTZ/UDLE (Deutsche Gesellschaft fur Technische Zusammenarbeit and Urban Development through Local Efforts), Kathmandu: GTZ.

Rabinow, P. 1992. "Studies in the Anthropology of Reason." *Anthropology Today* 8(5): 7–10.

Rademacher, A. 1998. "Restoration as Development: Urban Growth, River Restoration, and Riparian Settlements in the Upper Bagmati Basin, Kathmandu, Nepal." M.A. thesis, Yale University, School of Forestry and Environmental Studies.

———. 2005. "Culturing Urban Ecology: Development, Statemaking, and River Restoration in Kathmandu." Ph.D. diss., Yale University.

———. 2007. "A 'Chaos' Ecology: Democratization and Urban Environmental Decline in Kathmandu." In *Contentious Politics and Democratization in Nepal: The Maoist Insurgency, Identity Politics, and Social Mobilization since 1990*, ed. M. Lawoti, 299–321. Delhi: Sage.

———. 2008. "Fluid City, Solid State: Urban Environmental Territory in a State of Emergency, Kathmandu." *City and Society* 20(1): 105–29.

———. 2009. "When Is Housing an Environmental Problem? Reforming Informality in Kathmandu." *Current Anthropology* 50(4): 513–34.

———. 2010. "Restoration and Revival: Remembering the Bagmati Civilization," In *Symbolic Ecologies: Nature and Society in the Himalaya*, ed. A. Guneratne, 166–85. New York: Routledge.

Rademacher, A., and R. Patel. 2002. "Retelling Worlds of Poverty: Reflections on Transforming Participatory Research for a Global Narrative." In *Knowing Poverty: Critical Reflections on Participatory Research and Policy*, ed. K. Brock and R. McGee, 166–88. London: Earthscan.

Rademacher, A., and D. Tamang. 1993. *Democracy, Development, and NGOs in Nepal*. Kathmandu: Jeevan Support Press.

Raeper, W. and M. Hoftun. 1992. *Spring Awakening: An Account of the 1990 Revolution in Nepal*. Delhi: Viking.

Raffles, H. 2002. *In Amazonia*. Princeton: Princeton University Press.

Rai, H. 2002. "The Great Unifier." *Nepali Times* no. 76, January 11-17.

Rankin, K. 1999. "Kamaiya Practices in Western Nepal: Perspectives on Debt Bondage." In *Nepal: Tharu and Tarai Neighbors*, ed. Harold O. Skar, 27–46. Kathmandu: EMR.

Rao, V. 2006. "Slum as Theory: The South/Asian City and Globalization." *International Journal of Urban and Regional Research* 30:225–32.

Rebele, F. 1994. "Urban Ecology and Special Features of Urban Ecosystems." *Global Ecology and Biogeography Letters* 4:173–87.

Redclift, M., and T. Benton, eds. 1994. *Social Theory and the Global Environment*. London: Routledge.

Regmi, D. R. 1975. *Modern Nepal*. Calcutta: Firma K. L. Mukhopadhyay.

Regmi, M. 1978. *Thatched Huts and Stucco Palaces: Peasants and Landlords in 19th Century Nepal*. New Delhi: Vikas.

———. 1988. *An Economic History of Nepal, 1846–1901*. Varanasi: Nath.

Regmi, R. R. 1993. *Kathmandu, Patan, Bhaktapur: An Archaeological Anthropology of the Royal Cities of the Kathmandu Valley*. New Delhi: Jaipur.

"Reimagining Cities." 2008. Special issue. *Science* 319, February 8.

Riles, A. 2000. *The Network Inside Out*. Ann Arbor: University of Michigan Press.

Roy, A. 2004. *Urban Informality: Transnational Perspectives from the Middle East, Latin America, and South Asia*. New York: Lexington.

Roy, A., and N. AlSayyaad, eds. 2004. "The Gentleman's City: Urban Informality in the Calcutta of New Communism.," In *Urban Informality: Transnational Perspectives from the Middle East, Latin America, and South Asia*, ed. A. Roy and N. AlSayyad, 147–70. Lanham, Md.: Lexington.

Saberwal, V. K. 1999. *Pastoral Politics: Shepherds, Bureaucrats, and Conservation in the Western Himalaya*. Delhi: Oxford University Press.

Samuel, R. 1994. *Theatres of Memory: Past and Present in Contemporary Culture*. London: Verso.

Sassen, S. 1991. *The Global City: New York, London, Tokyo*. Princeton: Princeton University Press.

———. 1999. "Whose City Is It? Globalization and the Formation of New Claims." In *Cities and Citizenship*, ed. James Holston, 177–94. Durham: Duke University Press.

Scott, J. 1998. *Seeing Like a State: How Certain Schemes to Improve the Human Condition Have Failed*. New Haven: Yale University Press.

Seddon, D. 1987. *Nepal: A State of Poverty*. Delhi: Vikas.

Sen, A. 2002. "Globalization and Poverty." Transcript of lecture, Santa Clara (Calif.), University Institute on Globalization.

Sever, Adiran. 1993. *Nepal under the Ranas*. New Delhi: Oxford.

Shakya, S. M. 2001. "Environment." In *City Diagnostic Report for City Development*

Strategy, Kathmandu Metropolitan City. Kathmandu: Kathmandu Metropolitan Corporation (KMC)/World Bank.

Sharma, B. 1992. "Selling Dreams: Project Appraisal of the Kathmandu Urban Development Plans and Programs Study." *Himal* 5(1):34.

———. 2000. "My God! What a Big Drain!" *Kathmandu Post*, August 24.

Sharma, C. K. 1997. *River Systems of Nepal*. Calcutta: Sri K. K. Ray.

Sharma, S. 1997. "Present Water Quality Status of the Bagmati River." *Kathmandu Post*, June 8.

———. 2002. "The Hindu State and the State of Hinduism." In *State of Nepal*, ed. K. M. Dixit and S. Ramachandaran, 22–38. Lalitpur: Himal Books.

Shepherd, J. W. 1985. "Symbolic Space in Newar Culture." Ph.D. diss., University of Wisconsin, Madison.

Shrestha, N. 1990. *Landlessness and Migration in Nepal*. Boulder: Westview.

———. 1998. "Losing Shangri-La? The Environmental Degradation of Kathmandu." *Education about Asia* 3:11–18.

Shrestha, N., and P. Kaplan. 1982. "The Sukumbasi Movement in Nepal: The Fire from Below." *Journal of Contemporary Asia* 12:75–88.

Shrestha, R. R. n.d. *Water Pollution in Kathmandu Valley*. Report. Kathmandu: ENPHO.

———. 1999. "Application of Constructed Wetlands for Wastewater Treatment in Nepal." Ph.D. diss., Natural Sciences, University of Agricultural Sciences, Vienna.

Shrestha, R. R., and S. Sharma. 1996. "Trend of Degrading Water Quality of Bagmati River (1988–1995)." Paper presented to the Bagmati Environmental Seminar, March 11. Organized by Water Induced Disaster Prevention Technical Center/HMG Ministry of Water Resources.

Siu, H. 1989. *Agents and Victims in South China: Accomplices in Rural Revolution*. New Haven: Yale University Press.

———. 1990. "Recycling Tradition: Culture, History, and Political Economy in the Chrysanthemum Festivals of South China." *Comparative Studies in Society and History* 32(4): 765–94.

———. 2003. "Uncivil Urban Spaces in Post-Reform South China." Paper presented to the conference, "Gangs, Crowds, and Enclaves," Amsterdam, August 29–30.

Sivaramakrishnan, K. 1999. *Modern Forests: Statemaking and Environmental Change in Colonial Eastern India*. Stanford: Stanford University Press.

Sivaramakrishnan, K., and A. Agrawal. 2003. "Regional Modernities in Stories and Practices of Development." In *Regional Modernities: The Cultural Politics of Development in India*, 1–61. Stanford: Stanford University Press.

Slusser, M. 1982. *Nepal Mandala: A Cultural Study of the Kathmandu Valley*. Princeton: Princeton University Press.

Smart, A. 2001. "Unruly Places: Urban Governance and the Persistence of Illegality in Hong Kong's Urban Squatter Areas." *American Anthropologist* 103(1): 30–44.

Smart, A., and J. Smart. 2003. "Urbanization and the Global Perspective." *Annual Review of Anthropology* 32:263–85.

Smith, N. 1984. *Uneven Development: Nature, Capital, and the Production of Space.* Oxford: Blackwell.

Stanley International et al. 1994. *Bagmati Basin Water Management Strategy and Investment Plan: Final Report.* HMG/MHHP, World Bank/Japanese Grant Fund, April 1994, Kathmandu.

Stengers, I. 1993. *The Invention of Modern Science*, trans. D. Smith. Paris: La Decouverte.

Stiller, L. 1999. *Nepal: Growth of a Nation.* Kathmandu: Human Resources Development Research Centre.

Stoler, A. 2004. "Affective States." In *A Companion to the Anthropology of Politics*, ed. D. Nugent and J. Vincent. Malden, Mass.: Blackwell.

Strommen, K. 1991. *Planning for Urban Development and Environmental Upgrading: An Environmental Impact Assessment of the Bishnumati Link Road.* HMG Ministry of Housing and Physical Planning, and the University of Trondheim. Kathmandu: HMG.

Swyngedouw, E. 1996. "The City as a Hybrid: On Nature, Society, and Cyborg Urbanisation." *Capitalism, Nature, Socialism* 7(2).

———. 1999. "Modernity and Hybridity: Nature, Regeneracionismo, and the Production of the Spanish Waterscape, 1890–1930." *Annals of the Association of American Geographers* 89(3): 443–65.

———. 2004. *Social Power and the Urbanization of Water: Flows of Power.* New York: Oxford University Press.

———. 2006. "Metabolic Urbanization: The Making of Cyborg Cities." In *In the Nature of Cities: Urban Political Ecology and the Politics of Urban Metabolism*, ed. N. Heynen, M. Kaika, and E. Swngedouw, 21–40. London: Routledge.

Szeman, I. 1997. "Review of Arjun Appadurai's *Modernity at Large*." *Cultural Logic* 1(1): 1–6.

Tanaka, M. 1997. *Conditions of Low Income Squatters in Kathmandu.* Kathmandu: Lumanti.

Tarlo, Emma. 2002. *Unsettling Memories: Narratives of the Emergency in India.* Berkeley: University of California Press.

Taylor, P., and F. Buttel. 1992. "How Do We Know We Have Global Environmental Problems? Science and the Globalization of Environmental Discourse." *Geoforum* 23(3): 405–16.

Thapa, D. 2003. *Understanding the Maoist Movement of Nepal.* Kathmandu: Matin Chautari.

Thapa, D., and B. Sijapati. 2003. *A Kingdom under Siege: Nepal's Maoist Insurgency, 1996 to 2003.* Kathmandu: The Printhouse.

———. 2004. *A Kingdom under Siege: Nepal's Maoist Insurgency 1996 to 2003.* New York: Zed.

Thapa, K. 1994. "Upgradation of Squatter Settlements in Kathmandu." M.A. thesis, Tribhuvan University, Kathmandu.

Thapa, M. 2005. *Forget Kathmandu: an Elegy for Democracy*. New Dehli: Penguin, Viking.
Thompson, M., and M. Warburton. 1985. Uncertainty on a Himalayan Scale. *Mountain Research and Development* 5(2): 115–35.
Thompson, M., M. Warburton, and T. Hatley. 1986. *Uncertainty on a Himalayan Scale*. London: Ethnographica.
Tiwari, A. 1992. "Planning: Never without Aid." *Himal* 5(2): 8–10.
Tiwari, S. R. 1992. "No Future for an Urban Past." *Himal* 5(1): 5–7.
——— . 2001. *Nepal's Urbanization and Urban Poverty Alleviation: A Critical Look at the 10th Plan*. Draft paper for the MuAN Workshop.
Trouillot, M. R. 2001. "The Anthropology of the State in the Age of Globalization—Close Encounters of the Deceptive Kind." *Current Anthropology* 42:125–38.
Tsing, A. 2000. "The Global Situation." *Cultural Anthropology* 15:327–60.
——— . 2001. "Nature in the Making." In *New Directions in Anthropology and Environment*, ed. C. L. Crumley, 3–23. Lanham, Md.: Altamira Press.
——— . 2005. *Friction: An Ethnography of Global Connections*. Princeton: Princeton University Press.
Tuladhar, B. 1996. "Kathmandu's Garbage: Simple Solutions Going to Waste." *Studies in Nepali History and Society* 1(2): 365–93.
Udaya-Himalaya Network. 1992. "Aid in Nepal's Development: How Necessary?" Proceedings from a Panel Debate Organized by Udaya-Himalaya Network. Kathmandu.
UN-HABITAT. 2001. *The State of World Cities*. New York: United Nations Habitat.
——— . 2003. *The Challenge of Slums: Global Report on Human Settlements 2003*. New York: United Nations Habitat.
United Nations. 1996. *Habitat Agenda and Istanbul Declaration on Human Settlements Summary: Road Map to the Future*. United Nations Department of Public Information.
United Nations Population Division. 2003. *World Urbanization Prospects: The 2003 Revision*. New York: United Nations Population Division.
van Buuren, A., and J. Edelenbos. 2004. "Conflicting Knowledge. Why Is Joint Knowledge Production Such a Problem?" *Science and Public Policy* 31(4): 289–99.
Vandergeest, P., and N. Peluso. 1995. "Territorialization and State Power in Thailand." *Theory and Society* 24:385–426.
van der Veer, P. 1994. *Religious Nationalism: Hindus and Muslims in India*. Berkeley: University of California Press, 1994.
Varma, R. 2004. "Provincializing the Global City." *Social Text* 81(20): 4.
Walbridge, M. 1998. "Growing Interest in Urban Ecosystems." *Urban Ecosystems* 2(1): 3.
Werbner, R. P. 1984. The Manchester School in South-Central Africa. *Annual Review of Anthropology* 13:157–85.

West, P. 2005. "Translation, Value, and Space: Theorizing an Ethnographic and Engaged Environmental Anthropology." *American Anthropologist* 107(4): 632–42.

Whelpton, J. 1983. *Jang Bahadur in Europe: The First Nepalese Mission to the West.* Kathmandu: Shayogi Press.

———. 1991. *Kings, Soldiers, and Priests: Nepalese Politics and the Rise of Jung Bahadur Rana, 1830–1857.* New Delhi: Manohar.

———. 1992. *Kings, Soldiers, and Priests: Nepalese Politics 1830–1857.* Kathmandu: Ratna Pustak Bhandar.

———. 2005. *A History of Nepal.* New York: Cambridge University Press.

Williams, R. 1973. *The Country and the City.* New York: Oxford University Press.

———. 1980. "Ideas of Nature." In *Problems in Materialism and Culture: Selected Essays*, 67–85. London: NLB.

Worby, E. 1998. "Tyranny, Parody, and Ethnic Polarity: Ritual Engagements with the State in Northwestern Zimbabwe." *Journal of Southern African Studies* 24(3): 561–78.

———. 2000. "Introduction: Rethinking Zimbabwe's 'Crisis.'" In *Zimbabwe: The Politics of Crisis and the Crisis of Politics*, ed. Yuka Suzuki and Eric Worby. New Haven: Yale University.

———. 2001. "Grasping an Elusive State: Practical Epistemologies of Power in Zimbabwe in a Time of Crisis." Paper presented to the Yale University Program in Agrarian Studies, January 26.

World Resources Institute (WRI). 1996. *The Urban Environment.* New York: Oxford University Press.

Worster, D. 1990. "The Ecology of Order and Chaos." *Environmental History Review* 14:1–18.

———. 1994. *Nature's Economy: A History of Ecological Ideas.* London: Cambridge University Press.

———. 1996. "The Two Cultures Revisited." *Environment and History* 2(1): 3–14.

Wright, D. 2000 [1877]. *History of Nepal with an Introductory Sketch of the Country and People of Nepal*, trans. M. S. S. Singh and P. S. Gunanand. Delhi: Adarsh Enterprises.

Yami, H., and S. Mikesell. 1990. *The Issues of Squatter Settlements in Nepal: Proceedings of a National Seminar.* Kathmandu: Concerned Citizens Group of Nepal.

York, R., and E. A. Rose. 2003. "Key Challenges to Ecological Modernization Theory." *Organization and Environment* 16(3):273–88.

Zerner, C. 1993. "Through a Green Lens: The Construction of Customary Environmental Law and Community in Indonesia's Maluku Islands." *Law and Society Review* 28:1079–122.

———. 2003. "Moving Translations: Poetics, Performance, and Property in Malaysia and Indonesia." In *Culture and the Question of Rights: Forests, Coasts, and Seas in Southeast Asia*, ed. C. Zerner, 1–23. Durham: Duke University Press.

Zimmerer, K. 2000. "The Reworking of Conservation Geographies: Nonequilibrium Landscapes and Nature Society Hybrids." *Annals of the Association of American Geographers* 90(2): 356–69.

Zimmerer, K., and T. Bassett. 2003. "Future Directions in Political Ecology: Nature-Society Fusions and Scale of Interactions." In *Political Ecology: An Integrative Approach to Geography and Environment-Development Studies*, 274–95. New York: Guilford Press.

Index

Aaraati, 72
Abrams, Phillip, 187 n. 14
Activists, 26
ADB (Asian Development Bank), 14, 18, 143, 147
Administrative forgetting, 84
Agriculture, 67
Alley, Kelly, 72–73, 198 n. 17
Alliance for Religion and Conservation (ARC), 165, 171
Anglo-Gorkhali War, 192 n. 11
Appadurai, A., 39, 155
ARC (Alliance for Religion and Conservation), 165, 171
Arghajal, 3–4, 70
Arya cremation ghats, 10–11, 93, 111, 114
Asal Hindustan, 48, 103
Asian Development Bank (ADB), 14, 18, 143, 147
Authoritarian rule, 132–33
Auyero, J., 28, 58–59

Bagmati Area Sewage Construction and Rehabilitation Project (BASCRP), 112, 174
Bagmati Basin Management Strategy (BBMS), 14, 17–18, 60–61, 63, 144; as development policy, 74; prominence of, 62
Bagmati Basin Management Strategy and Investment Program, 60
Bagmati Bridge, 76, 86, 182; completion of, 3, 161–62; settlements below, 120
Bagmati Civilization (bagmati sabhyata), 83, 89, 128, 161, 181; remembering, 87
Bagmati River, 1, 42, 76, 176, 182; as crisis, 20–21; cultural respect for, 5; degraded reaches of, 12; drainage basin of, 17; as ethnographic anchor, 179–80; history of, 87; ideal form and function of, 7–8; path of, 17, 64, 198 n. 14; photo exhibition of, 170; pollution in, 2; predicament of, 145; sacredness of, 71, 73–74, 165; social life of, 65; source of, 66; stewardship of, 128; UN Park on, 157; urban reaches of, 33; water quality of, 67, 110–14, 167
Bagmati sabhyata. *See* Bagmati Civilization
Bahadur, Jung, 48
Bāhirako stories (outside stories), 109
Baidya, Huta Ram, 130, 201 n. 32; on democratic rights, 12; demonstration participation by, 2, 5; on development, 84–86, 180; ideas of, 128, 161–62; on Maitighar mandala, 123; personal memories of, 88; on restoration limits, 85; on river degradation, 86–87, 163; on river restoration, 89; as spokesperson, 83–84
Bansbari Tannery, 68, 70
Bansighat, 4–5
BASCRP (Bagmati Area Sewage Construction and Rehabilitation Project), 112, 174
BBMS. *See* Bagmati Basin Management Strategy
Beautification. *See* Emergency beautifications
Belonging, 33
Bhairavīs, 47
Bhairavs, 47
Bhaktapur, 35, 40, 47, 64, 191 n. 5
Bhatt, Nina, 102–3, 105
Bhim Sen Pillar, 161, 162
Billboards, as state domain, 118–19
Biophysical data, 15–16; actionable, 61
Biophysical degradation summary, 70

Biophysical processes, cultural processes and, 16
Birendra, King, 95–96; cremation of, 10–11; death of, 91; murder of, 55; popularity of, 54
Bishnumati Corridor, 69, 79
Bishnumati Corridor Environmental Improvement Project, 79, 142, 149
Bishnumati Link Road, 147, 208 n. 5
Bishnumati River, 1, 42, 128, 145; as crisis, 20–21; degraded reaches of, 12; drainage basin of, 17; as ethnographic anchor, 179–80; path of, 64; social life of, 65; source of, 66; UN Park on, 157; urban reaches of, 33
BOD statistics, 71
Brahmaputra River, xxii
Built environment, 35
Burch-Machlis model, 189 n. 26
Bureaucracy: growth of, 52; ineffectiveness of, 129

Campbell, Ben, 41
Cartographic space, 44
Caste (jāt), 45, 153 n. 254, 193 n. 14, 200 n. 24; assignment to, 48; eradication of, 50; state making and, 49
Cement, demand for, 37
Censorship, 108–10
Chandra Ghat, 4
Change, 8; spatial continuities and, 36
Chaudhary, Binod, 128
China, xxii, 64
Chromium contamination, 68, 70
Cities: environmental crisis of, 26; ethnic, 36; of global South, 22–23. *See also* Hypercity; Megacity
Citizens' groups, 128
Citizenship, 9, 25, 45
Civil Code (Muluki Ain), 48, 193 n. 14
Civil service, 52
Civil unrest, 117
Clashes, between government and rebels, 94
Class, 143, 153–54
Clean green ideology, 168
Collectivity, 33

Commission for the Resolution of the Landless Squatters Problem, 79
Communist Party Nepal-Maoist (CPN-M), 11–13, 94, 114; casualties of, 105; demands of, 97
Community Led Environmental Action Network, 68–69
Conflict, 13–15
Congress Party, 51
Consciousness, 87
Conservation, 41
Control, demonstrations of, 116
CPN-M. *See* Communist Party Nepal-Maoist
Crisis: of cities, 26; global narratives of, 22; of Kathmandu, 26, 29; in Nepal, 19; rivers in, 20–21; as usual, 21
Cultural diversity, 50
Cultural heritage activists, 15
Cultural history, 82–83, 88–89
Cultural processes, biophysical processes and, 16
Cultural restorationist narrative, 59, 83–89
Cultural rights, 4, 83, 89, 180

Dāphā, 2
Daurā suruwāl, 2
Davis, Mike, 141
Degradation, 70; causes of, 62; competing definitions of, 57–58; democracy as, 127–33. *See also* River degradation
Democracy: conditions of, 130; as degradation, 127–33; delay of, 51; disappointment and disillusionment with, 54–55, 179; environmental policy and, 87; expectations of, 8; experiments with, 50–51, 53–55, 105, 132, 179; forms of, 134; imagined, 127; river degradation and, 130–31
Democratic polity, constitutional monarch of, 50–51
Democratic processes, change and, 8
Demonstrations, 2, 55
Deśa, 46–47

Deśa ghumne, 48
Deuba, Sher Bahadur, 97, 105
Development, 7, 19, 37, 84–86, 180; BBMS and, 74; cultural history, modernization and, 88; as employment source, 54; internal, 112; in Panchayat period, 51; political opposition and, 52; problems of, 20, 77, 178–79; socioeconomic, 52; of Tinkune, 120–21, 205 n. 6
Developmentalism, 34, 72, 185 n. 3
Developmentalist logics of morality, 34
Development experts of river restoration, 15
Dhunge dharas, 122
Dipendra, Prince, 96, 202 n. 5; death of, 11, 91, 93–95; as murder suspect, 92
Diversion tunnel, 110–11, 204 n. 20
Dixit, Kanak Mani, 167; on Kathmandu cultural degradation, 19–20
Domestic wastewater, 69
Doubt, truth and, 110
Dry season, 66–67
Dysfunctionality, 19

Earth systems, 28
Ecological affinities, 38. *See also* Environmental affinities
Ecological degradation, 12
Ecological illegitimacy, 144
Ecological improvement, 144–45
Ecological logics, 178
Ecological utility, 160
Ecologies of reform, 152–54
Ecology: limits of, 27; moral logics of, 139; moral power of, 181; power of, 38; in practice, 16, 177, 180; scientific dimensions of, 182; social practice and, 13; state making and, 176. *See also* Urban ecology
Ecosystem change, scientific characterizations of, 16
Effective symbolic authority, 107
Emergency beautifications, 116–17, 125–27, 132, 136–37

Emergency period, 45, 55, 105, 107; aftermath of, 149; civil unrest following, 108, 117; initial reaction to, 119, 155; narrative coherence of, 106; settlements during, 145–49; urban life in, 118
Emergency powers, of king, 51
Encroachment, 6, 131
Engaged universal, 29, 183
Environment, 30, 35, 179; affinities of, 178; change in, 15, 120–26; integrity of, 33; problems of, 26–27; protection laws for, 128; rehabilitation of, 151–52. *See also* Ecological affinities
Environmentalism, state making and, 21
Environmentalist collectives, 178
Ethnicity, 40
Ethnic political parties, 40
Ethnographic perspective, 38
Evictions, 123–24, 126, 147
Exchange, 27–28
Exclusion, forms of, 177–78
Expatriate community, 164–65

Facility maintenance, 113
Fecal coliform bacteria, 67, 71
Fieldwork, 14, 146, 157
First Five-Year Plan, 52
Fish ladder, 3, 76
Flash flooding, 82
Forgetting, 84
Formal activities, 141
Fratricidal massacre, 95
Friends of the Bagmati, 14, 168–69, 172; goals of, 165–67, 174; international ties for, 173; language used in, 170–71; Nepali members of, 166; strategy of, 166
Fund designation, 53

Gandagi, 73
Ganga, 73
Ganges River, 17
Ghats, 4
Global and local conundrum, 39, 155–56, 170–71
Global city, 25

Global positionality, 156, 209 n. 2
Global South, 22–23, 29, 188 n. 21
Gopal Mandir, 85, 162
Gorkha, 9
Gorkhali-Shah era, 47
Governance, 26, 137; forms of, 56; opaque, 121
Government censorship, 108–10
Guhyeswori plant, 113
Gunakamadeva, 43
Gutschow, Neils, 34–36
Gyanendra, King: family of, 92–93; holdings of, 134; legitimacy of, 118; opinions of, 101, 202 n. 4; Parliament reinstatement by, 14; power consolidation by, 148; as regent, 92; respect for, 100; rule of, 13, 55, 133; suspicion of, 11

Habitat restoration, 80
Hallā (rumor), 109
Hāmro Bāgmatī, 170–72
Herzfeld, Michael, 151, 207 n. 1
Hierarchy, implications for, 35
Himalayan deforestation, 20
Hinduism: ideology, morality and, 36; Hindu kingdom and, 9; in Nepal, 9, 54; ritual sites of, 72
His Majesty's Government (HMG), 14, 60, 111–12
Historical perspective, 38
Hjortshoj, Keith, 38
HMG (His Majesty's Government), 14, 60, 111–12
Homeownership opportunities, 149
Housing, 151; concrete, 37; demand for, 18; supplies for, 77. *See also* Informal housing
Housing advocates, 15, 82–83, 124, 139–40, 201 n. 31; on legitimacy, 143; on municipal policy, 148–49
Housing-focused narrative, 59, 75–83
Hukumbāsī, 142
Human actors, moral logics of, 139
Human agency, 16, 28
Human encroachment, 18, 62
Human-nature interactions, 27, 30
Human social agents, 27
Hypercity, 22, 188 n. 17. *See also* Cities

Identity: national, 52; Nepali, 49–50, 59; Newar, 40; river-focused, 155
India, xxii, 64; Pakistan and, 118
Indian, as outsider, 142
Indra, 43
Indra Jatra festival, 48
Industrial waste dumping, 57
Informal activities, 141; green rehabilitation of, 151
Informal housing, 42, 152–54, 208 n. 3; proliferation of, 24–25; urban environment and, 140–41. *See also* Sukumbasi settlements
Informal sector, 141
Information confirmation, 108–9
Infrastructure projects, 85
INGOs (International nongovernmental organizations), 53
Inside order, 34
International development projects, river degradation and, 3, 63
International nongovernmental organizations (INGOs), 53
International visitors, 165
Intractability, 19
Irrigation, small-scale, 66

Jal, 70
Jana andolan (*janā āndolan*), 31, 54, 104, 107, 137. *See also* People's Movement
Jana andolan II, 133
Jan yuddha. *See* People's War
Jāt. *See* Caste
Juddha Ghat, 4
Jyotir Lingam, 9

Kalmochan Temple, 3
Kalopul area cleanup, 168
Kamaiyas, 135–38, 139, 178
Kanteswara devata, 43
Kantipur, 43
Karuwas, 119
Kathmandu, 17, 47, 64; centrality of, 122–23; civic life, 166; conquest of,

47; as convergence point, 43; cultural degradation of, 19–20; environmental crisis of, 26, 29; environment of, 18–19, 30; as ethnic city, 36; expatriate community of, 164–65; growth in, 77; history of, 71; municipal water supply for, 66; origin of, 43; urban growth rates of, 22; urban planning of, 163
Kathmandu Urban Development Project (KUDP), 63, 66
Kathmandu Urban Development Project Bishnumati Corridor Environmental Improvement Program, 14
Kathmandu Valley: cultural and natural history of, 84; mandalic patterns across, 35; Newars and, 40; organization of, 8; Slusser on, 36
King: authority of, 11; deposition of, 117; divine power of, 102–3; emergence of, 102; as symbol, 10
Kirtipur, 150
Knowing, ways of, 26, 90
Knowledge, 26, 189 n. 25; production of, 28; scientific, 28–29; social, 182; urban ecology as, 32, 140
Koirala, Girija, 97, 128
KUDP (Kathmandu Urban Development Project), 63, 66
Kumari, Padma, 129

Lalitpur, 64
Laws of Manu, 48
Levy, Robert, 34, 36
Liberation ecologies, 180
Lichhavi Dynasty, 46
Liechty, Mark, 48
Local participation, promotion of, 168–69, 170–72
Local positionality, 156, 209 n. 2
Lohani, Anu, 131
Lokeswara, 43
Loktantra period, 45, 134, 153, 207 n. 23
Low, Setha, 190 n. 34
Lower Bagmati Basin, 17
Lumanti, 14, 209 n. 10; changed terms for, 148; commitment of, 151; conference of, 80; funding of, 201 n. 30; site visit by, 150; on sukumbasi, 144–46

Mahendra, King, 5, 11, 49, 51
Mainali, A., 129
Maitighar mandala, 45, 120–24, 135–37
Malla era, 45, 47, 164, 191 n. 4
Manchester School, 25
Mandala, 34–36, 38, 44–46, 122, 138, 205 n. 4. *See also* Maitighar mandala
Maobadi, 94, 146
Maoist Movement, 39
Maoist state, 127
Marcus, G., 33
Marginalized groups, 139
Massacre, 114, 145, 202 n. 5; aftermath of, 105; confusion about, 93–94; as conspiracy, 95, 97; meaning assigned to, 100; narrative of, 95; official investigation of, 96, 100
Meaning, 159; production of, 28
Megacity, 22, 188 n. 17. *See also* Cities
Memorial pillar, 158–61, 174
Middle Bagmati Basin, 17
Migrants, 15, 152; fate of, 149; lack of plan for, 75; marginalization of, 150; narrative constructions of, 141; positive impacts of, 81–82. *See also* Riparian
Migration trends, 75–76; control of, 125; internal, 142–43; narrative of, 131
Mini Nepal, 157–58
Ministry of Housing and Physical Planning complex, 157
Modern global citizenship, 173
Modernity, 34, 137, 174; democratic, 104–5
Modernization, 19, 151, 153; cultural history and development and, 88; of Nepal, 20; tradition and, 37
Monarchy, 8–9, 102–4, 138; abolishment of, 175; future of, 133; legitimacy of, 12; opposition to, 55; power of, 50
Moral authority, 101, 103

Moral geography, 84, 201 n. 34, 207 n. 2
Morality: developmentalist logics of, 34; Hindu conceptions of, 36
Moral logics, of ecology, 139
Moral social order, 33
Moser, Caroline, 141
Muluki Ain (Civil Code), 48, 193 n. 14

Narratives, 196 n. 1; importance of, 164
National disorder, symbolic marker of, 2
National Habitat Committee, 145–46
National history, form and content of, 46
National identity, 52
Nationalism, 50, 195 n. 19
National narrative promotion, 163
National unification, 9, 50
Nation building, under Mahendra, 49
Nation-state, 15, 127, 195 n. 19
Natural order, political order and, 179
Nature, 16; biophysical accounts of, 26; production of, 30, 204 n. 2
Neighborhood watch system, 169–70
Ne Muni, 43
Nepal, xxii, 64; autonomy of, 112–13; conflict in, 13; crises in, 19; cultural unity of, 49; developmentalism in, 34; divisions within, 97; foreign assistance to, 52; free elections in, 31; as Hindu kingdom, 54; Hindu sites in, 9; history of, 45–46; identity of people of, 49–50, 59; modernization of, 20; political life in, 39; remaking of, 135; rural revolution of, 152; state of, 8–9; as strategically important, 52; UN contributions of, 2; urban development in, 31
Nepālīpan, 49
Nepālmandala, 45, 190 n. 1; first date reference to, 46; reshaping of, 48
Nepal River Conservation Trust, 167
Nepal's United People's Front, 94
Nepal Tourism Board, 167
Newar: favoring of, 8; identity, 40, 99; Kathmandu Valley and, 40; mandala and, 35–36; middle-class, 99

News gathering, 108
Nongovernmental organizations (NGOs), 53

Objective study, as impossible, 41
Official state-development narrative, 59, 62–75
Order, demonstrations of, 116
Outside order, 34

Pakistan, India and, 118
Panchayat: constitution, 51; democracy, 51; period, 10, 50, 51, 53, 107
Park banaune, 119
Parliament, 175; democracy of, 8; reinstatement of, 14, 55, 133
Pashupati Area Development Trust, 111–12, 203 n. 19
Pashupatinath, 9–11, 93, 111, 131, 186 n. 6
Patan, 40, 47
Peace talks, 98, 105
People's Movement (jana andolan), 2, 31, 54
People's War (jan yuddha), 11, 13, 96, 116, 165, 206 n. 13; associations with, 132; intensification of, 19, 94
Pigg, Stacy, 53–54
Piped water supply contamination, 67
Politics: cartographic space and, 44; of engagement, 39; environmental change and, 15; natural order and order of, 179; power of, 46, 63; uncertainty of, 132–33; urban environmental space and aspirations of, 117, 177
Polity, mandala space and, 44–46
Pollutants, 69–70
Population debates, 25
Poverty, 22, 76–77
Power, 26, 33; active, 46; autonomous, 112; concentration of, 122, 148, 193 n. 13; decentralization of, 104; of ecology, 38; to enact environmental change, 44; global spheres of, 173; inversion of relations of, 45–46; of kings, 102–3; monarchical, 50, 102–4; political, 46, 63; state, 41; symbolic, 46; urban ecology and, 33
Prachanda, 98, 105, 175

Pradhan, Jyoti, 99
Prajatantra, 54, 107
Production: of knowledge, 28; of locality, 155–56; of meaning, 28; of nature, 30, 204 n. 2; of social knowledge, 182; of urban ecology, 32–33
Public demonstrations, 2
Puja, 9, 72
Purity, 112

Radio Nepal, 49
Rajaraniko photo, 103
Rana, Jung Bahadur, 88, 186 n. 1, 201 n. 37
Rana isolationism, 48, 50
Rao, Vyjayanthi, 23
Realm purity, 48
Rebel army, 94
Reforestation initiatives, 20
Refugees, 146
Relative consumption patterns, 25
Remembering history, 84
Renaming, 133
Resettlement schemes, 143
Resettlement site, 150–51
Restoration goals, 61; competing logics of, 144; limits to, 85
Riles, Annelise, 33
Ring Road, 98, 175
Riparian: migrant settlements, 7; revegetation, 82; settlers, 5; zone, 83, 140. *See also* Migrants; Sukumbasi settlements
Ritual ablutions, 72; obstacles to, 132
Riverbank migrant settlement, 5. *See also* Sukumbasi settlements
River degradation, 15–18, 113, 176, 187 n. 10; Baidya on, 86–87, 163; engaging and defining, 43, 89; international development projects and, 3, 63; local aspects of, 26; morphological change associated with, 156; obstacles created by, 132; perceptions of, 7, 21, 58; politics and, 61, 74, 129–131; pollution and, 57; reversal of, 60; settlement and, 139; urban ecology and, 30

River restoration efforts, 13, 26, 63, 66; Baidya on, 89; criticism of, 18; of experts, 15; full scale, 112; meaning of, 17; official ideas of, 143; stalemate of, 126
River restoration politics, 6, 46, 177; conflict and, 14
Rivers, 71–72, 172–73; conflicts and, 6; flow quantity, 57, 62, 74; focused actors, 173, 177, 182; identity and, 155; management of, 3, 180; morphology, 57, 68, 197 n. 8; narratives, 59; pollution, 57; religious practice centered on, 15; stewardship, 163, 169; water quality of, 57, 62, 82
Riverscape: as clean, 151; importance of, 171–72; logics of belonging to, 181; monuments, 156; narratives, 180; practices, 87–88; production of, 16; respect for, 131
RNA (Royal Nepal Army), 13, 105
Road construction, 5
Royal family, 10–11, 93, 111
Royal massacre. *See* Massacre
Royal Nepal Army (RNA), 13, 105
Rural agricultural land, saturation of, 76
Rural dissent, 146
Rural-to-urban migrants, 75, 77–78, 125, 139, 152, 205 n. 13. *See also* Sukumbasi

SAARC (South Asia Association for Regional Cooperation), 116, 118–20, 204 n. 1
Sacred buildings, distribution of, 35
Sacred land of Hindus, 47
Sacred landscape studies, 34
Samāchār, 49
Sand: demand for, 37; deposition of, 69; mining of, 18, 62, 68–69, 87, 143
Satdobato, 99, 175
Save the Bagmati Campaign, 14
Science, 26–27
Scientific knowledge, 28–29
Sediment loads, 69, 198 n. 15
Self-centered attitude, 129–30

Sewage: discharge of, 17–18, 57, 62, 65, 144; odor of, 6; treatment, 81, 111, 113
Shah, Prithvi Narayan, 192 n. 7; conquest of, 45, 47; pride in, 103; on realm purity, 48; statue of, 134
Shah Dynasty, 8, 47–48
Shakti, 47
Sharma, Khagendra, 136
Sharma, Prayag Raj, 50
Sharma, Sudhindra, 194 n. 16, 199 n. 19, 205 n. 8
Shiva, 47
Shiva temples, 9
Shrestha, Laxman, 147, 158–61, 163, 174
Shrestha, Nanda, 53
Shrestha, R. R., 68, 197 n. 9
Sivaramakrishnan, 16
Slum settlements, 24. *See also* Sukumbasi settlements
Slusser, Mary Shepherd, 34, 186 n. 6, 191 n. 6; on Kathmandu Valley, 36; on mandalic patterns, 35, 44–45; on river sacredness, 72
Snaan, 72
Social cohesion, 15, 178
Social constructionist approaches, 30
Social forgetting, 84
Sociality, forms of, 174, 178
Social knowledge production, 182
Social life, cartographic space and, 44
Social order, Levy on, 36
Social organizations: implications for, 35; registration of, 23–24
Social practice, ecology and, 13
Social rehabilitation, 151–52
Social Service National Coordination Act, 23
Social Services National Coordination Council (SSNCC), 23–24, 53
Socioeconomic development, 52
Socioenvironmental exclusion, 58
Socioenvironmental inclusion, 58
Solid waste dumping, 62, 65, 69
Solid Waste Management and Mobilization Center (SWMMC), 69
South Asia Association for Regional Cooperation (SAARC), 116, 118–20, 204 n. 1
Sovereignty, 136–38
Spatial continuities, change and, 36
Spatial form, 36
Spatial pattern, 34
Squatter settlements, 24, 142
Śraddhā, 70
SSNCC (Social Services National Coordination Council), 23–24, 53
State experts on river restoration, 15
State making, 186 n. 8; caste and, 49; ecology and, 176; environmentalism and, 21; urban ecology and, 15, 117–20
State power, 41
Stupa, 120
Sukul mats, 4
Sukumbasi, 4, 80–82, 144–48, 153; environmentalism and, 178; hostility toward, 7; as term, 77–78, 141–42, 185 n. 2. *See also* Rural-to-urban migrants
Sukumbasi settlements, 5, 65, 81, 199 n. 23; characterization of, 142; diversity in, 200 n. 24; growth of, 77, 139; as obstacle, 113, 147; official gestures toward securing, 124; question of, 24–25; raids on, 147; reforming, 151; removal of, 123–24, 126, 143–44; suspicion of, 154; upgrading, 79–80. *See also* Informal housing
Sundarighat, 68
Sundarijal, 68, 167
Swistun, D. A., 28, 58–59
SWMMC (Solid Waste Management and Mobilization Center), 69

Teku, 1, 7, 64; cleanup, 169; convergence of, 65; Kalopul area of, 168
Teku Dovan, 17, 42–43, 66
Teku-Thapatali Project, 66
Templescape, formation of, 47
Territorial claims, 5
Terrorist and Disruptive Activities Ordinance, 105–6
Thapa, Bhim Sen, 162, 210 n. 9

Thapa, Manjushree, 93
Thapathali settlement eviction, 126, 147
Tharu laborers, 135–38
Til Ganga, 167
Tinkune development, 120–21, 205 n. 6
Tirthas, 88
Tiwari, Sudarshan Raj, 75–78, 151–52
Trade wealth, 47
Tradition, 37
Transnational capitalism, urban processes and, 25
Transnational circulations, 155
Tribhuvan, King, 49–51
Truth, doubt and, 110
Tsing, Anna, 29
Tuladhar, Padma Ratna, 97

UN (United Nations), 2
UN Conference on Human Settlements, 80
United Nations (UN), 2
Universal human rights, 75
Universality, 15
Universals, 29, 183
UN Park, 3, 66, 126, 156–61, 206 n. 14
Unplanned space, erasure of, 123
Upako vanegu, 48
Upper Bagmati Basin, 17, 64, 66, 69, 196 n. 5
Urban, rethinking of, 22
Urban beautification campaign, 116
Urban communities: depopulation of, 139; as ecosystem, 27; growth of, 18, 22, 78, 145–46, 181, 199 n. 20; planet as, 183; processes of, 25; waste produced by, 42. *See also* Urbanization

Urban Community Support Fund, 149
Urban construction activity, 68
Urban ecology, 115; affective dimensions of, 164; approach to, 16; aspects of, 79; decline of, 8, 183; experiences of, 183; housing advocacy framing of, 75; as knowledge, 32, 140; mandala and, 45; meaning of, 29–41, 61, 89, 173; multiple, 28–30; official, 62–75; postdemocratic, 125–27; power and, 33; in practice, 38, 161, 177–78, 180; production of, 32–33; questions about, 31, 176; river degradation and, 30; shift towards, 23; social life of, 26–29, 178; state making and, 15, 117–20
Urban environmental space: informal housing and, 140–41; political aspiration and, 117, 177
Urban informality, 141
Urbanization, 22–23, 25, 29
Urban nature, 12–17

Vāmshavalī, 43, 102
Vastu Shastras, 36
Vishnu, king as reincarnation of, 102
Vishnuko amsh, 102

War on Terrorism, U.S., 118
Waste: of contemporary urban life, 42; dumping of, 62, 65, 69
Water, 67, 196 n. 6; sacred value of, 72
Weir dams, 143
Whelpton, J., 34
Williams, Raymond, 154
World Water Day, 171

ANNE M. RADEMACHER
IS AN ASSISTANT PROFESSOR
IN THE ENVIRONMENTAL STUDIES
AND METROPOLITAN STUDIES PROGRAMS
IN THE DEPARTMENT OF SOCIAL AND
CULTURAL ANALYSIS AT NEW YORK
UNIVERSITY.

Library of Congress Cataloging-in-Publication Data
Rademacher, Anne.
Reigning the river : urban ecologies and political
transformation in Kathmandu / Anne M. Rademacher ;
foreword by Dianne Rocheleau.
p. cm. — (New ecologies for the twenty-first century)
Includes bibliographical references and index.
ISBN 978-0-8223-5062-0 (cloth : alk. paper)
ISBN 978-0-8223-5080-4 (pbk. : alk. paper)
1. Kathmandu (Nepal) — Environmental conditions.
2. Kathmandu (Nepal) — Civilization.
3. Kathmandu (Nepal) — Politics and government.
4. Urban ecology (Sociology) — Nepal — Kathmandu.
5. Stream ecology — Nepal — Kathmandu.
I. Title. II. Series: New ecologies for the twenty-first century.
DS495.8.K3R33 2011
954.96 — dc23   2011021959